CUMBRIA L'

KT-151-732

3800305060299 4

'It's SO ...
the h...
Sarah ..., author of *The Essex Serpent*

'Smart plotting, strong characters and a
deft exploration of complicity and
betrayal add up to a stand-out read.'
Guardian

'Not just a tense psychological drama but
a meditation on the complex view... Dive in.'
Susie Steiner, author of *Persons Unknown*

'A taut, cleverly constructed novel about what we see,
know and understand – and how they are not always
the same thing. Prepare to step into the darkness...'
Linda Green, author of *After I've Gone*

'A clever thriller that examines provoking issues.'
Good Housekeeping

'Kelly is a beautiful writer who can conjure an atmosphere
with exceptional skill... a fantastic psychological thriller
replete with mystery and misdirection.'
Alex Marwood, author of *The Darkest Secret*

'Twisted, atmospheric and perfectly plotted. Her best yet.'
Jenny Blackhurst, author of *The Foster Child*

'A superbly dark and complex psychological thriller
from a really exceptional author. Kelly's beautiful
and atmospheric writing pulls you in from the first page.'
Kate Hamer, author of *The Doll Funeral*

'I devoured it . . . weaves such a delicate line between everyone's
version of the truth, it was impossible to stop reading.'
Holly Seddon, author of *Don't Close Your Eyes*

'Multi-layered and stylish, with characters you feel you
know, dilemmas you can imagine grappling with, and a
hurtling plot that's unpatronising and clever to the last page.'
Helen Fitzgerald, author of *Viral*

'Gripping and atmospheric.'
Heat

'A masterclass in drip-fed suspense, beautifully
written and astutely observed. This book left me
extremely tense throughout – in a good way!'
Susi Holliday, author of *The Deaths of December*

'An addictive read. The mystery and majesty of the
total solar eclipses lends an eerie inevitability
to the doomed and twisted attempts of Erin Kelly's
all-too-human characters to control their lives.'
Isabelle Grey, author of *The Bad Mother*

'Brilliantly plotted, packed with suspense.'
Women & Home

'A haunting and beautifully twisted tale of
desire and deceit. Erin Kelly is the kind of writer
whose work you read and instantly wish it were your own.'
Colette McBeth, author of *The Life I Left Behind*

'*He Said/She Said* is dazzling ... you simply won't
be able to look away. Erin Kelly is a class act on top form.'
Lucy Dawson, author of *You Sent Me a Letter*

'Beautifully written, gripping and as twisty
and dangerous as a 400-page-long
rattlesnake, this is Erin Kelly at the absolute
top of her game, and everyone else's too.'
William Ryan, author of *The Constant Soldier*

'Brilliant. A masterclass in how to write
a gripping psychological thriller.'
Louise Millar

'Love and its twisted opposite, truth and
lies that shade from little white to pitch-dark.
Here be images to haunt you and characters
to make you doubt your own judgement.'
Lucie Whitehouse, author of *Keep You Close*

'Thought-provoking and sensitively handled.'
Stylist

'Deliciously tense and twisty.'
Sharon Bolton, author of *The Craftsman*

'A clever, compulsive read, which sinks its
claws into you and won't let go.'
Red magazine

'Erin Kelly keeps you guessing to the very last page,
she cleverly reminds us of our own fallibility and gullibility.'
Adele Parks, author of *The Image of You*

'It's magnificent. Stunningly twisty plot and
weep-makingly brilliant writing'
Marian Keyes, author of *The Break*

'Gradually, with skilful pacing and growing tension,
Erin Kelly reveals shocks and twists that
go on until the very last page.'
The Times

'Erin Kelly speaks the language of tension fluently.
He Said/She Said will make you question everything
you know and suspect everyone you think you can trust.'
Janet Ellis, author of *The Butcher's Hook*

'Expertly constructed.'
The Sunday Times

'Superbly twisty and readable thriller; pure enjoyment.'
Jenny Colgan, author of *The Endless Beach*

Erin Kelly

He Said/She Said

HODDER

First published in Great Britain in 2017 by Hodder & Stoughton
An Hachette UK company

This paperback edition published in 2018

1

Copyright © E S Moylan Ltd 2017

The right of Erin Kelly to be identified as the Author of
the Work has been asserted by her in accordance with
the Copyright, Designs and Patents Act 1988.

All rights reserved. No part of this publication may be reproduced, stored
in a retrieval system, or transmitted, in any form or by any means without
the prior written permission of the publisher, nor be otherwise circulated
in any form of binding or cover other than that in which it is published and
without a similar condition being imposed on the subsequent purchaser.

A CIP catalogue record for this title is available from the British Library

Paperback ISBN 978 1 444 79714 5
eBook ISBN 978 1 444 79717 6

Typeset in Sabon MT by Hewer Text UK Ltd, Edinburgh
Printed and bound in Great Britain by Clays Ltd, St Ives plc

Hodder & Stoughton policy is to use papers that are natural, renewable
and recyclable products and made from wood grown in sustainable
forests. The logging and manufacturing processes are expected to
conform to the environmental regulations of the country of origin.

Hodder & Stoughton Ltd
Carmelite House
50 Victoria Embankment
London EC4Y 0DZ

www.hodder.co.uk

A total eclipse of the sun has five stages.

First contact: The moon's shadow becomes visible over the sun's disc. The sun looks as if a bite has been taken from it.

Second contact: Almost the entire sun is covered by the moon. The last of the sun's light leaks into the gaps between the moon's craters, making the overlying planets look like a diamond ring.

Totality: The moon completely covers the sun. This is the most dramatic and eerie stage of a total solar eclipse. The sky darkens, temperatures fall and birds and animals often go quiet.

Third contact: The moon's shadow starts moving away and the sun reappears.

Fourth contact: The moon stops overlapping the sun. The eclipse is over.

We stand side by side in front of the speckled mirror. Our reflections avoid eye contact. Like me, she's wearing black and like mine, her clothes have clearly been chosen with care and respect. Neither of us is on trial, or not officially, but we both know that in cases like this, it's always the woman who is judged.

The cubicles behind us are empty, the doors ajar. This counts as privacy in court. The witness box is not the only place where you need to watch every word.

I clear my throat and the sound bounces off the tiled walls, which replicate the perfect acoustics of the lobby in miniature. Everything echoes here. The corridors ring with the institutional clatter of doors opening and closing, case files too heavy to carry wheeled around on squeaking trolleys. High ceilings catch your words and throw them back in different shapes.

Court, with its sweeping spaces and oversized rooms, plays tricks of scale. It's deliberate, designed to remind you of your own insignificance in relation to the might of the criminal justice machine, to dampen down the dangerous, glowing power of the sworn spoken word.

Time and money are distorted, too. Justice swallows gold; to secure a man's liberty costs of tens of thousands of pounds. In the public gallery, Sally Balcombe wears jewellery worth the price of a small London flat. Even the leather on the judge's chair stinks of money. You can almost smell it from here.

But the toilets, as everywhere, are great levellers. Here in the ladies' lavatory the flush is still broken and the dispenser has still run out of soap, and the locks on the doors still don't work properly. Inefficient cisterns dribble noisily, making discreet speech impossible. If I wanted to say anything, I'd have to shout.

In the mirror, I look her up and down. Her shift dress hides her curves. I've got my hair, the bright long hair that was the first thing Kit loved about me, the hair that he said he could see in the dark, pulled into a schoolmarm's bun at the nape of my neck. We both look . . . demure, I suppose is the word, although no one has ever described me that way before. We are unrecognisable as the girls from the festival: the girls who painted our bodies and faces gold to whirl and howl under the moon. Those girls are gone, both dead in their different ways.

A heavy door slams outside, making us both jump. She's as nervous as I am, I realise. At last our reflections lock eyes, each silently asking the other the questions too big – too dangerous – to voice.

How did it come to this?

How did we get here?

How will it end?

First Contact

I

London is the most light-polluted city in Britain, but even here in the northern suburbs, you can still see the stars at four o'clock in the morning. The lights are off in our attic study, and I don't need Kit's telescope to see Venus; a crescent moon wears the pale blue planet like an earring.

The city is at my back; the view from here is over suburban rooftops and dominated by Alexandra Palace. By day it's a Victorian monstrosity of cast iron, brick and glass, but in the small hours it's a spike in the sky, its radio mast tipped with a glowing red dot. A night bus of the same colour sweeps through the otherwise empty park road. This part of London has a truer 24-hour culture than the West End. No sooner does the last Turkish kebab shop shut than the Polish bakery takes its first delivery. I didn't choose to live here, but I love it now. There is anonymity in bustle.

Two aeroplanes blink across each other's paths. One floor below me, Kit is deep in sleep. He's the one going away, yet I'm wide awake with pre-trip nerves. It is a long time since I slept through the night but my wakefulness now has nothing to do with the babies in my belly who tapdance on my bladder and kick me awake. Kit once described real life as the boring bit between eclipses but I think of it as the safe time. Beth has crossed the world to find us twice. We are only visible when we travel. A couple of years ago, I hired a private detective and challenged him to find us using only the paper trail of our previous lives. He couldn't trace us. And if he couldn't do it, then no one else can. Certainly not Beth, and not even a man of Jamie's resources. It has been fourteen years since one of his letters found me.

This total eclipse will be the first Kit has seen without me since he was a teenager. Even the eclipses he had to miss, he missed with me, because of me. It's not a good idea to travel in my condition, and I'm so grateful to be in this condition that I don't begrudge missing the experience, although I am terrified for Kit. Beth knows me. She knows *us*. She knows that to hurt him is to destroy me.

I watch the moon set in its slow arc. Following its course is a deliberate act of mindfulness, the living-in-the-moment therapy that is supposed to stop my panic attacks before they can take hold. The telltale early symptom is there; a subtle standing-to-attention of all the tiny hairs on my skin, a feeling that some-one's trailing a gossamer scarf over my forearms. They call it somatising, a physical manifestation of psychological damage. Mindfulness is supposed to help me separate the soma from the psyche. I play join-the-dots with the constellations. There is Orion, one of the few constellations everyone can identify, and, flung a little to the north, the Seven Sisters that give the nearby neighbourhood its name.

I rock back and forth from the heels to the balls of my feet, concentrating on the carpet fibres under my bare toes. I can't let Kit see me anxious. In the short term, it would ruin his trip, and after that, he would suggest more psychotherapy, and I've taken that as far as I can. There's only so far you can get when you're holding on to a secret like mine. The psychotherapists always say that the sessions are confidential, like their Ikea couch is a sacred confessional. But my confession is a broken law, and I can't trust anyone with it. There is no statute of limitations for what I did in this country, and none in my heart.

When my breathing evens out, I turn away from the window. There is just enough light to see Kit's map. Not the original of course, that was destroyed, but a painstaking recreation of it. It's a huge relief map of the world, crisscrossed with curves of red and golden thread, measured to the nearest millimetre, glued down with characteristic precision. The gold arcs mark the eclipses he has already seen; the red those we can expect to see in our lifetimes. Part of the ritual is coming home after a trip to replace red threads with gold. (Being Kit, he has calculated his life expectancy using family history, lifestyle and longevity trends, and allowed for infirmity curtailing travel when he's ninety. So we should see our last eclipse in 2066.)

Years ago, Beth trailed her fingers over the first map and that's when I told her about our plans.

I wonder where on the planet she is now. Sometimes I wonder if she's even still alive. I have never wished her dead – for all that she put us through, she was a victim too – but I have often wished that she could be . . . deleted, I suppose, is the right word. There's no way of finding out. Try to look up 'Elizabeth Taylor' and see how far you get without the actor or the novelist making a nonsense of your search. Using the diminutive 'Beth' does little to narrow it down. She seems to have vanished as effectively as we have.

I haven't looked Jamie up for years. It's too uncomfortable, after my part in it all. His public relations crusade paid off and these days when you search his name the crime comes up but only in his preferred context. The first few hits are about his campaigning work, the support he gives to wrongly accused men and rightly accused men too, calling for anonymity up to the point of conviction. I can never get beyond the first few lines before I start to feel sick. I still need to keep myself informed, so I got around the problem by setting up a Google alert that links his name to the only word that matters. There's no point combining his name with Beth's in a search; her lifelong anonymity is guaranteed. That's the law whatever the outcome of this kind of trial. I suppose she was lucky – we all were, in a way – that the case pre-dated social media and the keyboard vigilantes whose blood sport is identification.

Light on the landing tells me Kit's awake. I take a deep breath in and a longer breath out and I am calm. I have beaten this attack. I roll up the sleeves of the sweater I'm wearing. It's Kit's, and it doesn't do me any favours, but it fits and I seem to have been at the stage where I dress for comfort for years now. Even before I conceived, the steroids gave me hips and breasts for the first time in my life, and I still haven't worked out how to dress around curves.

I pad down the stairs, edging past the flat-packed cots on the landing. When Kit comes home we'll have to convert Juno and Piper's room at the back of the house into a nursery. Superstition, a reluctance to do anything until he has survived this trip, has held me back.

I find him sitting up in bed, already checking his phone for the weather report, his pale copper hair at mad angles. The words *don't go* try to punch their way out of my mouth. Knowing he would stay if I asked him to is all the reason I need to let him leave.

2

KIT
18 March 2015

I lie awake for a few seconds, listening to Laura's footsteps overhead, and savouring the Christmas-morning feeling. The thrill never lessens when the abstract numbers on the calendar finally take shape into days. I have known for years that on 20th March 2015, the moon will block the sun from view, making a black disc in the sky. Total eclipses of the sun have been dots on the timeline of my life since I first stood beneath the moon's shadow. Chile 1991 was *the* eclipse of the last century; seven minutes and twenty-one seconds of pure totality. I was twelve years old and I knew that I would devote the rest of my life to recapturing the experience. Nothing compares to witnessing a total solar eclipse under a cloudless sky. Until I met Laura, it was the closest I came to understanding religion.

The sheets on her side of the bed are cold. When she comes in, her belly entering the room a beat before she does, her cheeks are sunken from tiredness. Her hair is tied up, the roots showing, a millimetre of brown that looks black against the platinum lengths. She's wearing one of my old sweaters, pushed up to the elbows. She has never looked lovelier. I had worried, when we first started trying for a baby, whether I'd miss that ectomorph gawkiness I always loved, but there's a new pride at seeing Laura's body change because there's something of me in there.

'Get back into bed,' I say. 'It's not good for you to be leaping around.'

'Ah, I'm awake now. I'll go back to bed when you've gone.'

In the shower I run through today's itinerary one last time, the finer details in my grand plan. I'll catch the 05.26 from Turnpike Lane Tube, then the 06.30 from King's Cross to Newcastle, where I will meet Richard at 09.42. From there a chartered minibus will take us to Newcastle docks and at a pleasingly round 11.00 we will board the *Princess Celeste*, a 600-berth cruise ship that will take us across the North Sea, past Scotland and halfway to Iceland, where the Faroe Islands lie. Most of Friday's eclipse will be over water, but even a calm sea is never still and the best photography is done on land. I had to choose between the Faroes or Svalbard, north of the Arctic Circle. (It was Laura who wanted me to go to the Faroes. The biggest crowds will be in Tórshavn, on Stremoy, the largest island, and she believes in safety in numbers.) In two days' time, at 8.29 a.m., the moon will start to creep across the surface of the sun and slowly build to two and a half minutes of total eclipse.

I towel-dry the beard Laura insisted I grow for the trip, then dress carefully in the clothes I laid out the night before. My work clothes – not a uniform, but they might as well be – hang neatly in the wardrobe, tugging at my conscience. Delighted as I am at the prospect of five days away from the optical lab, I can't help but feel guilty at taking annual leave to travel when I could have tacked it on to my paternity leave. Then I think about the chemicals I've been breathing in for so long that they line my lungs, and the stiff neck that's been craned over lenses all year finally hinging upwards to look at the sky, and I think sod it. I've got the rest of my life to play the provident father. What's five days, in the wider scheme of things?

I put on a long-sleeved thermal vest, and then over that my lucky T-shirt, a souvenir from my first eclipse. It says Chile '91 on it – countries always claim the eclipse as their own, even when the shadow falls over three continents – and is in the colours of the

Chilean flag. A crude black circle in its centre represents the covered sun, surrounded by the flares of a corona. When my dad bought it from a roadside hawker it was virtually a dress on me. Mac refused to wear his but I wouldn't take mine off even to wash it. It fits me now but it won't in a few years unless I follow Mac's lead to the gym. There's a burn on the collar where Mac flicked a lit joint at me during an argument in Aruba, in 1998. On top of these layers goes the magnificent finishing touch, a work of art in chunky black and white wool. Richard and I bought matching Faroese jumpers online months ago. We're stamping down hard into our carbon footprints by taking them home to the country where the sheep grazed and where the wool was spun and knitted.

I check my phone again, in case weather conditions have changed in the last ten minutes, but the forecasts remain gloomy. There's a thick blanket of cloud across the whole archipelago. 'Eclipse chasing' sounds like a misnomer, and I've learned to defend the term over the years. How can you chase a phenomenon when you're the one moving, and the phenomenon is standing still? First of all, there's nothing still about an eclipse; the darkness comes at more than a thousand miles an hour. Well, it's true that there's no changing the co-ordinates. The shadow will fall where the shadow will fall, in a pattern that was established when we were still primordial soup. But clouds are not nearly so predictable. An unanticipated cumulus can disappoint a crowd of thousands who only moments before were standing confidently in sunshine. The thrill is in outwitting the weather. My fondest memory of my father is from Brazil '94, Mac and me riding loose in the back of Dad's VW, speeding along a pot-holed highway until we found a patch of blue sky. (He was, in retrospect, drunk behind the wheel; I try not to dwell on that.)

These days, naturally, there are apps. Breaks in the cloud can be pinpointed with much greater precision, and it's not unusual for entire coach parties not to know their viewing destination until five minutes before first contact. I turn my phone face down. I will go mad if I think too hard about the weather. Fortunately I've always been good at shutting out thoughts that would distract or upset me. In the moments when I allow myself to think about the past, which is not often – it only gets shoved to the forefront of my consciousness when there's an eclipse on the horizon, and Laura's triggers go off – in those rare moments, it seems that life since the Lizard has been lived as though under a malfunctioning neon light. A subtle but constant vibrating strobe that you learn to live with, even though you know that one day it will trigger some kind of seizure or aneurysm.

The smell of fresh coffee wafts up the stairs. Laura's in the kitchen, five steps down and at the rear of the house. Our scrubby little back garden is in darkness. She has filled a mug for me and she's wrapping a sandwich in foil. I kiss her behind her right ear and inhale the buttery scent of her. 'Finally, the subservient house-wife I've always wanted. I should leave you on your own more often.' I feel the skin on her neck tighten as she smiles.

'It's the hormones,' she says. 'Don't get used to it.'

'Promise me you'll go back to bed once I've gone,' I say.

'Promise,' she says, but I know Laura. I had hoped that preg-nancy might slow her down but if anything the steroids have sped her up, so she'll power through the day until collapsing in a heap somewhere around 9 p.m. She sweeps the worktop clean with a sponge and puts the empty coffee pods in the bin. With her back to me, she performs a tiny act, meaningless to anyone but me, that twists at my guts. She swipes at her bare forearms, twice, as if brushing imaginary cobwebs from her skin. It is months, if not years, since I have seen her do this and it always means she's thinking about Beth. I wish for the millionth time that she had my discipline when it comes to the past, or rather the way the past might impact our future. Why waste energy anticipating something that could never happen? She gets like this with every eclipse, even though it's been nine years since Beth's last known movements. She turns around with a too-wide smile, literally putting on her brave face for me. She doesn't know I saw her brush at her arms. She might not even know she did it.

'What've you got planned for today?' I ask her, to gauge her mood as much as anything.

'Calling a client first thing,' she says. 'And then this afternoon I thought I'd tackle my VAT. You got anything planned?'

I take heart from her joke. When she's about to crash, her sense of humour is the first thing to go.

My rucksack has been packed for three days now. Half the considerable weight is camera equipment, lenses, chargers and my tripod, batteries and waterproofs, and then spares of everything. The camera is in its own bag, too precious to leave unattended in a luggage rack. My phone goes into the breast pocket of my orange windcheater.

'Very chic,' says Laura drily. 'Have you got everything you need?' I put the sandwich in my other pocket, check my Oyster card is easily accessible and then hoist on the rucksack. I nearly fall backwards under its weight.

Without warning, Laura's smile drops and she brushes her forearms, twice in succession. This time we make eye contact and denial is as pointless as explanation. Reassurance is all I can give.

'I've checked the passenger records,' I say. 'There's no Beth Taylor on the list. No Taylors. No Elizabeth anything. No B or E anything, female.'

'You know that's completely meaningless.'

Indeed I do. Laura thinks that Beth has changed her name. I disagree; it's a reflection of Laura's paranoia. With a name like that, you can hide in plain sight. That was, after all, the inspiration behind our own rebranding. Why hide a needle in a haystack when you can hide a strand of hay? 'And even if it's true,' presses Laura, 'all that means is she's not on your ship. What if she's on the ground?'

I speak deliberately slowly. 'If she is there, she'll be looking for a festival. Somewhere there's a sound system and a load of bongos, that's where she'll expect to find us. I'm going to be travelling with a load of retired Americans. And even if she doesn't, Tórshavn's a big place that'll be crawling with tourists, eleven thousand people.' I smooth down my beard. 'There's my cunning disguise. I'll be on the lookout. I'll be walking around with a periscope, checking all the corners before I go anywhere.' I mime peeping through my fingers; she doesn't laugh. 'Mac's round the corner, Ling's two streets away, my mum's an hour away, your dad's on the phone whenever you need him.'

'I can't help it, Kit.' I can see she hates herself for crying by the way she bites down hard on her lip. I draw her against me and with my other arm, I shake her hair from its messy topknot and comb it with my fingers, the way she likes me to. A tear rolls off the waterproof surface of my jacket. I take a deep breath and say the only thing she needs to hear.

'If you want me to stay, I'll stay.'

She pulls out of the hug and, for a horrible moment, I think this is to let me take my rucksack off. But instead she gets my camera bag and hangs the strap around my neck, solemnly, like she's awarding an Olympic medal. This is how she gives me her blessing and I can see what it takes.

'Look after yourself,' she says.

'You look after yourself. Yourselves,' I correct, and without thinking of the consequences, I kneel down to kiss her belly. My thighs scream with the effort of standing upright again.

'It could be worse,' I say. 'I could be going to Svalbard. Someone got mauled by a polar bear in Svalbard just last week.'

'Heh,' she says, but her heart's not in it. To her, Beth Taylor is scarier than any flesh-eating bear. I know what she's thinking; the first time Beth lashed out in retribution, she told us herself that she only stopped because they caught her. She actually admitted it would have been far worse if she'd attacked the person, rather than the property.

Outside, dawn has yet to break and the street glows orange in patches. There are two stone steps from our front door to street level. From the pavement, I turn to look up at Laura, who's rolled her sleeves down over her wrists, her hands cupping her bump. I have what Mac would call a moment of clarity. I'm about to leave my pregnant, over-medicated, anxious wife to travel across the seas to another country where there is every chance the woman who nearly destroyed us will be waiting for me.

'I'm not going,' I say, and I'm not calling her bluff. Laura frowns back.

'You bloody well are,' she says. 'Over a grand this trip's cost. Go on.' She shoos me off down the street. 'Have the time of your life. Take some pictures. Come back with beautiful stories for our babies.'

I take a last look down at my feet; the pavements here are treacherous enough without throwing an undone bootlace into the mix. 'The chances of her finding me are tiny,' I say, but Laura has already closed the door, and I realise that I was talking to myself anyway.

It's a five-minute walk from our house on Wilbraham Road to Turnpike Lane Station, less if you cut through Harringay Passage, a useful if rather Dickensian passageway that halves our grid of streets. I cross Duckett's Common, looping around the swings and slides where our friends' children play. Broken glass crunches under my feet.

Sweat is pouring off me already, cooling in my beard. For all the salt on my lips, it's the lie that sits bitter on my tongue. There's no way I could possibly have checked the passenger lists. That sort of thing is basic data protection. I can't believe Laura didn't pick up on it. When her anxiety's active, she gets super-powers of perception. Paranoia alerts her to even the tiniest shift in my body language and she picks up the slightest dilution of the truth.

I only ever keep things from her that I know will upset her.

Turnpike Lane Tube Station is still closed when I get there, its art deco splendour undermined by crappy shop hoardings and peeling billboards. At precisely 5.20 a.m. the iron lattice gates are pulled wide by a TfL worker in a royal blue fleece. The only other passenger is a tired-looking black woman in a tabard, probably off to clean some office in the city.

I glide down the escalator, lost in thought. It seems unlikely that Beth will be on my ship, but not impossible that she will be somewhere on the Faroes. I'm glad to be travelling alone, and that I don't have to think about Laura's safety. I have been protecting my wife against the fall-out of what happened on Lizard Point for so long. I will do anything to keep it that way.

3

LAURA
10 August 1999

The National Express coach was stationary on the A303 outside Stonehenge. It seemed like half the world was travelling into the West Country for the eclipse. The sky was the same grey as the standing stones, the ancient clock on the soft green hill. If I had to be stuck in traffic, this seemed like an appropriate place; people don't realise that Stonehenge was once used to predict eclipses as well as mark midsummer. But after over an hour staring at the sacred site, even I was struggling to remain awestruck.

Every time the weather report came on to the coach driver's radio, a rake-thin man with a straggly druidic beard sitting near the front would stand up, clap his hands and give us the update. The chances were we would be clouded out. My fellow travellers to the festival in Cornwall mostly whooped and cheered anyway, in a younger, cooler version of the famous British stoicism that had seen our grandparents through the Blitz and our parents through caravanning holidays. For them, it seemed, the eclipse was just an excuse for a festival; a bonus if they witnessed it, but if they didn't there would still be the music. Kit cared deeply about the eclipse, and I knew there would be a corresponding gloom in his mood.

He, along with Mac and Ling, had already been on the festival site for two days, setting up the tea stall that would hopefully turn a small profit. I hadn't eaten since my breakfast meeting with the man from the recruitment agency, and I'd changed in the toilets at Victoria Coach Station. The clothes I'd worn for my job interview were stashed in my rucksack. I kicked my army boots, pressing them down as though on to an accelerator, and wondered if I'd get to Lizard Point before nightfall.

Eventually the coach squeezed through the bottleneck, which was caused not by roadworks but drivers rubbernecking at the debris of a pile-up. Soon Wiltshire gave way to the chalk horses of Dorset. By lunchtime we were in Somerset. The chemical toilet got blocked somewhere in Devon. When we entered Cornwall, a genuine cheer went up. The chimneys of disused tin mines seemed to sprout from the hills almost as soon as we crossed the border, and here and there the county standard, the distinctive black flag with its white cross, fluttered proudly. I felt the press of the sea on either side as England petered into a peninsula and the familiar weight building inside me to know that on the southernmost point of the county, Kit was waiting for me.

We had been together six months at that stage. That time was less a honeymoon period and more a fugue state. It should have derailed our university finals, but Kit reaped the rewards of a lifetime of study and a photographic memory, and I fluked it with a question about the one text I'd studied and a ready supply of amphetamine sulphate. Kit insists it was love at first sight; I think it took about twelve hours. We agree to differ.

Ling and I were in our third year at King's College London when she started going out with a media studies student called Mac McCall (even his mum didn't call him by his real name, Jonathan). I liked Mac, up to a point – he was good looking in a russety sort of way, funny and exciting and generous with his drugs, but he had a way of taking over whatever space he was in, and I resented him slightly for crashing into my friendship with Ling. I was in no hurry to meet his twin brother, who was studying theoretical astrophysics at Oxford. Chalk and cheese, I thought, and I was right. Mac is your classic extrovert – he draws his energy from people, from crowds – while Kit is a textbook introvert. Conversation drains him; ideas recharge him.

Eclipses brought us together, in a way. As a very young woman I chased any experience that purported to be authentic or alternative to the mainstream culture I used to sneer at. I only liked grimy clubs and right-on bands no one had heard of, and I went out with a lot of boys who looked like Jesus. I thought that standing in a field watching a star disappear would be the ultimate climax to the ultimate rave, a special effect beyond the imagination and budget of any club promoter. When Ling said that she and Mac had found a way to see the upcoming total eclipse in Cornwall and get paid for it, I was in.

Mac lived in Kennington, in an ex-council flat with low ceilings and walls covered in swirling fluorescent fractal posters. I walked in across a forest floor of torn-up Rizla packets. The bulb in the living room had blown, and the place was lit by candles in jam jars. Kit, down from Oxford for the weekend, was a coiled figure in a shadowy corner, his face hidden behind a floppy strawberry-blond fringe, a woolly black jumper pulled down over his wrists. He seemed paler than Mac, in all ways.

'Dearly beloved,' began Mac, his hands busy with a lump of hash and a lighter (he could talk and roll a joint the way most of us can talk and blink). 'We are gathered here today to find a way we can go to a festival without actually having to pay for it. The best mark-up I can find is on hot drinks, teas and coffees, and if we work in shifts, we should turn a tidy profit.' Mac was surprisingly entrepreneurial for a self-professed anarchist. He wore Amnesty T-shirts and preached peace and love but only to those who mirrored his own values. He made a peace sign by way of a greeting, but thought nothing of keeping his neighbours awake all night with deafening techno.

'Right,' he said, sparking the joint. The lighter's flare showed me Kit's angles for a second: brows straight as rulers, an arrow-head nose above a set mouth. 'There are about ten festivals in the

West Country that week. They're all still in the planning stages, but I've got as much information together as I can, to help us decide which one fits best with our ethos.'

I tried to catch Ling's eye to share a smile at Mac's pomposity but she was gazing in rapt adoration. I felt the usual sting of exclusion.

'The big eclipse festival is in Turkey, but that's way beyond our budget,' said Mac. 'Plus, how often does this come around on home turf?'

'Less than once in a lifetime,' Kit piped up from his corner. His voice was Home Counties, educated: Mac's without the mockney drawl. 'A total eclipse needs *really* precise alignment. It's hard to average, but the last one here was in 1927 and the next one won't be until 2090. And we didn't have a single total eclipse between 1724 and 1925.'

'All right, Rain Man,' said Mac, going back to his list. He discounted three festivals where the music was 'too mainstream', and another where the sponsor was 'too corporate.' Ling, who had the predicted visitor numbers, ruled out a tiny gathering that wouldn't be worth our while. We were left with one festival in North Devon, and another on the Lizard peninsula in Cornwall. 'It's too close to call,' said Ling.

'Bro?' said Mac. Kit got to his feet without using his hands. He's taller than me, I thought. Measuring men against my own five-nine frame was often the first sign I had that I was attracted to someone. From a leaning plywood bookcase with half the shelves missing, he produced a sheaf of computer printouts.

'The thing about Cornwall, all of the West Country really, is that there are a handful of micro-climates. The weather conditions really can vary mile by mile. So I've correlated average sunlight and rainfall with all the festivals and plotted this against the path of totality. By my reckoning, this location gives us our best chance of seeing the sun.' He unfolded a battered Ordnance Survey map of Cornwall, and pointed at the Lizard peninsula.

'The Lizard Point Festival it is,' said Mac, and Kit's smile went from tentative to broad. 'I think this calls for a celebration.'

The celebration consisted of a bottle of Jack Daniel's, passed around while Mac played DJ and Kit shuffled his papers. I was used to Mac and Ling's public displays of affection, and I assumed that Kit would be too, but when they started snogging on the sofa, he was clearly mortified, blushing scarlet and eyes looking anywhere but in my direction. After a while he disappeared into the kitchen. I cleared my throat loudly.

'Sorry,' said Mac, smoothing down his T-shirt. 'We'll go next door.'

'How am I supposed to get home?' It was a long, dark walk back to our little flat in Stockwell and the last bus had gone. I hadn't drunk so much that I was willing to risk the walk, and back then it wouldn't have occurred to me to get a cab.

'Kit'll walk you,' Ling said, getting unsteadily to her feet. Her bra was already unhooked. She winked at me over her shoulder. 'Don't shag him, though. It'll make things really awkward in Cornwall.'

If I hadn't already been entertaining the idea, I'd have decided to shag Kit just to spite her.

'*Oh*,' he said on returning to find me there alone, then retreated to his corner where he sat cross-legged, drumming his fingers in perfect time to the music.

'That's really clever, what you did with those charts,' I said eventually, to break the silence.

'It's just maths,' he shrugged, but his fingers stilled.

'I really struggled with maths,' I said. 'In secondary school, I had this geometry teacher who was drawing shapes on the board, and she paused and clutched her bosom and said, "Of course the most beautiful shape of all is the circle," and I felt locked out of the secret of it. Of the story of it.'

Kit tilted his head to one side, as if he could read me better on the oblique. 'That's better than what most people say,' he said. 'There's a kind of pride in being crap at maths, an inverse snobbery about it, such a lack of *respect*. I don't know if it's a defence mechanism or what, but it drives me crazy. They don't realise how beautiful maths is. Like, listen to this tune.' I tried to give the music my full attention, but it was difficult with the bed next door squeaking on the offbeats.

'They've been together, what, six months now,' Kit said, his eyes travelling towards the wall where the sounds were coming from. 'He'd better not cock this one up like he usually does.'

My head suddenly cleared. 'Hang on, what?' Ling and I were used to going into battle for each other. 'Is he doing the dirty on her?'

'God, no!' said Kit, back-pedalling clumsily. If Mac had the gift of charm, poor Kit had barely grasped the tenets of tact. 'It's just. He hasn't got the best track record. You know. With. Girls. *Women*. But I'm sure this is fine. Ling.' He put the bottle to his lips and tilted, clearly dismayed to find it empty.

'I can see who stole all the moral fibre in the womb,' I said, to put him at his ease.

'Hardly. Mac's the one who goes on all the marches and stuff.'

'That's what he wants the world to see. Don't you think it's more important how you treat the person next to you?' I asked.

In Kit's answering smile, I saw a quiet integrity, so different to the boys who'd come before him, their politics printed on their T-shirts, and just as changeable.

'Well, I . . .' Whatever he was going to say next was interrupted by a low growl from the room next door that could have come from either one of them.

'Anyway,' I said, desperate to cover the noise, 'You were going to tell me what this tune had to do with maths.'

Kit took the cue to turn up the music. A sitar riff danced around a thumping bass. His brow was furrowed in concentration. 'Leibniz said, Music is the mind counting without being conscious it's counting. An eclipse is maths; it's the most beautiful maths there is.' Lost for words in the face of such intensity, I made what I hoped was an encouraging face. 'The moon, right, it's one-four-hundredth the diameter of the sun, but it's four hundred times closer to earth, so it looks like they're the same size.'

I got the feeling I was going to need some kind of animated diagram if this was to make sense, but it seemed important not to look ignorant in front of him. 'How many eclipses have you seen?' I said, to bring the conversation, if not down to earth, closer to my orbit, and he was off. He told me about driving around the Americas with his dad, about the time in India where they watched the sun vanish with his dad, his brother 'and a load of really confused goats' edging their way along the wall of a ruined temple. He told me about Aruba, where they'd stood on sand so hot it melted plastic and seen Venus and Jupiter, 'clear and round as pins in a corkboard'. How the planets and stars always came out and stopped hiding, like they didn't want to miss it. 'When you see it, when you stand underneath it, it isn't science. All that falls away.' The colour rose in his cheeks as he got technical again, talking me through the stages of an eclipse, describing the flaming ring of fire called the corona appearing around the sun, how the 1919 eclipse provided evidence for Einstein's theory of relativity by showing that the sun's mass bent light from distant stars. Partly I was listening to what he said and interested, but also partly I was watching him talk; the way his face changed completely when he was animated, the way his eyes skittered about in shyness as well as recall. I tried to imagine Mac talking for this long about any subject other than himself, and the thought made me smile. 'Ah, I'm boring you,' he said.

'You're really not.'

'Mac says I go on too much. What about you? You're on Ling's course, aren't you? What are you going to do after graduation?'

I told him about my grand plan, to work in the City for a few years until I had the CV to defect to the charity sector. I'd seen too many of my dad's bumbling, earnest friends shaking collection tins, spending all day chasing a handful of coppers.

'There's only one way of making a difference to people's lives, and that's with money. And if you want money, you've got to go where there's loads of it.'

'Like Robin Hood, but with spreadsheets and hedge-fund managers?'

'That's a very good way of putting it.'

As the candles burned to stubs, we exchanged potted biographies, the way you do when you're young and all you've got to offer beyond record collections and your degree course are the people you grew up with. There was a sense, that night with Kit, that this information was important; here's what you're getting into, we were saying. Still want to go for it?

I learned that Kit's parents, Adele and Lachlan, lived in Bedfordshire, in the third house in as many years, downsizing first when Lachlan had lost his job, and again when he'd drunk away their remaining equity. Adele was teaching textiles at a sixth-form college while she waited for her husband to die. Lachlan McCall, Kit said, had been what they call a functioning alcoholic, then a jobless one, until one day, a couple of years earlier, his liver had finally given out. They wouldn't put him on the transplant list until he stopped drinking. And he was still on the bottle.

'Mac's never said anything,' I said.

'Well, he wouldn't, would he? You've seen the way he is. I mean, I like a drink now and then, but he's on another level. I don't even think losing Dad will stop him.'

His lip quivered once. When I offered up the loss of my mum, Kit simply said, 'Oh, Laura, I'm so sorry. That's no age for grief.'

The floor between us suddenly contained two graves, one full and overgrown, one empty and waiting. I became aware of the background music and for a long time neither of us said anything. When the CD whirred to its end, Kit gulped a couple of times, as if he was working up to a big speech, before mumbling into his jumper, 'I like your hair.'

(*I like your hair*, or some variation on those words, was the first thing most people said to me back then. It had been waist-length mousy-brown string when I arrived at university; desperate to reinvent myself, I'd bleached it in my halls-of-residence bathroom on my first night away from home, turning it into a skein of bright white silk. I've worn it that way ever since, doing the roots every three weeks. It makes me sound incredibly high-maintenance, but I don't wear much make-up and I don't follow fashion. When you only have one vanity, I think you're allowed to indulge it.)

Kit reached over to pick up a strand; it looked luminous in the candlelight. 'I could never lose you in a crowd, even in the dark,' he said. When he put his hand to my cheek, I could feel his heart-beat in his palm.

We had fumbled, disappointing sex by the dim light and feeble heat of a two-bar electric fire. It was nerves that ruined it; nerves and the unspoken mutual knowledge of how much it already mattered. But January nights are long, and by the morning, apprehension had worn off and something new had taken over. I felt swept clean by Kit, rewritten, unable to think I had been with anyone else. We never had that conversation. I joined the dots between his anecdotes and worked out that before me, his love life had been a series of false starts. And if he was doing the same with me – extrapolating the data, as he would have said, from my own carefully pared stories – well then, he must have known that nothing else had come close to what we had. From his own stories, I soon understood that no one outside his family had ever noticed Kit much unless it was for passing an exam, and felt sorry for everyone who'd overlooked him or hadn't tried to get beyond his clumsy exterior. They were missing out on a whole world. That he let me in was an honour, and a point of pride; I took the responsibility I felt for his heart seriously and vowed every night to live up to his image of perfection.

Only a very young woman would think this way.

The longed-for *I love you* came in different words, spoken in Kit's bed in Oxford, in the middle of the night.

'Laura.' My own name broke urgently into my sleep. '*Laura*.'

'What's happened? What's wrong?' I tried to search his face in a shaft of weak light from the landing, but caught only an unreadable silhouette. His fingers threaded themselves through mine, as if to prevent escape.

'I'm sorry, I couldn't sleep. I need to know.' He sounded on the edge of tears as he took my hot hands in his cold ones. 'This. *Us*. Is it the same for you as it is for me? Because if it isn't . . .' He was shaking. I finished the sentence for him in my head. Because if it isn't, I don't think I can handle it. Because if it isn't, end it now. I wanted to laugh at the simple beauty of it, but could tell how much courage it had taken him to ask.

'It's the same for me,' I said. 'I promise. It's the same.'

That conversation was our marriage proposal. From the following day we talked unselfconsciously in terms of 'when we're married,' of our future children, the house we'd live in when we were old, and when Kit spoke of eclipses he would travel to ten, twenty, thirty years in the future, it was taken for granted that I would be there too, holding his hand under the shadow.

4

LAURA

18 March 2015

A peachy dawn breaks gently over Alexandra Palace, a graceful backdrop to my VAT return. Keeping my PC offline, I fill in a spreadsheet in the study, grateful for the distracting humdrum logic of the task. The paranoia of last night hasn't faded with the dark. If anything I'm getting worse, the closer Kit gets to boarding. It's one of those days when I wish I worked in an office, so that I could flush out anxiety with small talk about last night's telly or whose turn it is to get the teabags in. Instead, it's just me and a red telephone that seems to glow with menace.

A couple of weeks ago, I dropped my guard at a conference and was caught in a publicity photograph. The women's refuge I sometimes work for posed with their sponsors holding one of those giant novelty cheques. Since I closed the deal, I was there in the background. The refuge has put the picture on their website, and I need to ask them to take it down, or to crop me out or even Photoshop me out. At least they haven't printed my name. Kit and I decided when social media was in its infancy that we would leave no digital footprint. In the days where you can find anyone at the click of a mouse, we have to work harder than ever to make ourselves untraceable. I do what I always do when I'm faced with a phone call I don't want to make, and write a list of what I want to say, refining it into bullet points. When I'm training new fundraisers, I tell them that the single most important thing – even more important than believing in your cause – is to have a script. Never make a phone call without one. If you can't condense your pitch into four bullet points, you'll never hit your targets. It never usually fails me, but now I stall after the first point.

- *I cannot have my picture published on the internet.*

I heard on Radio 4 last year that you can buy facial recognition software that means all someone has to do is upload your photograph – scans count – and the app will trawl online images until it finds a match. It sounded to me like something from one of Kit's beloved science-fiction novels, but so once did all the technology we now take for granted. We know that Beth has at least one photograph of us and – not knowing at first how sly she could be – we had so many snapshots just lying around the flat. She could have had any one of them copied and replaced before we had noticed. I must be one of the only women who *wants* crows' feet and jowls but Kit says I've aged well. I don't know if it's flattery or just the fact that we've barely spent a night apart in fifteen years, so he can't see the changes: a hollowing under the eyes; the slanting accents, grave and acute, carved in the skin between my eyebrows. Or maybe he can and he's being kind.

It is only eight thirty, still before office hours, and I realise there's a cowardly way around this. I call the refuge knowing I'll go straight to voicemail, leave a message asking them to take my picture down for personal reasons and hope that they're too embarrassed to dig deeper. I'm lucky that I make a good living doing something I love and believe in, but my career has definitely been impeded by my reluctance to publicise myself along with the causes I raise money for. I still get head-hunted once or twice a year but my answer is always the same. I cannot have a high profile.

I knew from early on there was madness in the heat of Beth's moments. It wasn't until Zambia that I understood she was as dogged as Jamie in her own way. I often wonder if she lives, like I do, with our history bubbling constantly in the background, spilling over only when an eclipse is coming. You couldn't live at that level for the best part of fifteen years. It must come in waves, as it does for me. Or as it must for Jamie, whose campaign is not governed by alignment of the planets but legal mechanics.

After hours in the chair, I'm stiff all over and when I stand up my lower back cramps in response. I use the loo for the fourth time this morning, then rearrange the magazines in the bathroom into his-n-hers piles: *New Scientist*, *New Humanist* and *The Sky at Night* for Kit; *New Statesman*, *The Fundraiser* and *Pregnancy and Birth* for me. For balance I take the stairs crabwise, straightening the pictures on the walls as I go. It's a series of eclipse shots, glossy black circles surrounded by tongues of white fire that look more like abstract art than anything from nature. They are in chronological order, deliberately unlabelled, although even if I were to mix them up Kit would be able to tell you exactly when and where each one was taken.

On the console table by the front door sits our wedding photograph in a little silver frame. It's a bittersweet image; two frightened kids wearing other people's clothes on the steps of Lambeth Town Hall. Kit's bandages had only come off the day before.

There's a thudding noise as the builders next door start work for the day. Until a few years ago the house next door to the left had two different families crammed into it; last year, it was bought by Ronni and Sean, who are now converting the flats back into a house big enough for their three children. Like everyone else who moves in these days, they are furious at having been priced out of Crouch End. Our neighbourhood is known as the Harringay Ladder, because on the map the streets look like eighteen rungs strung between Wightman Road and Green Lanes. Wilbraham Road is the sixth rung down. When we told Ronni and Sean we'd been on the Ladder since 2001, Sean whistled and said, 'You must be *swilling* in equity.' Once, perhaps, if everything had gone according to plan, but Kit's earning power isn't what we thought it would be and maintaining Edwardian houses doesn't come cheap. If we hadn't had the roof replaced we'd be able to see the stars from our bed whether we wanted to or not. And that's before you count the IVF. After the third failed round it was clear that the only way forward was a hefty remortgage.

Kit hates Ronni for something she said to me a few weeks later. She was hugely pregnant with a toddler in a pushchair, and as I helped her up the steps to her front door, she said, 'You must really rattle around in there, with no kids. We should swap! Our flat's just about the right size for two.'

I kept it together until she was through her door, then I ran home, crashing so hard into Kit that I had his toothmark indented in my forehead for the rest of that day. I threw myself on the sofa and wailed while Kit called her a rude, clumsy, insensitive bitch and threatened to go next door and say something. (He's much tougher on my behalf than he ever is on his own.) I had to beg him not to.

I've packed an emergency bag in the hallway, my maternity notes wedged into the side pocket. Everyone, from my consultant to my mother-in-law, says I won't need it, but not to have the bag prepared is to tempt fate. I'm not nervous about the birth. I'm booked in for a C-section at thirty-seven weeks. What really worries me is having three new relationships thrust on me overnight; a mother twice over and a co-parent. I suppose I can't see how it will work, sharing Kit. It's always been just me and one other person – me and my mum, then me and Dad, a succession of intimate friends throughout school, then me and Ling and now me and Kit. I suppose that, for a while, Beth was virtually living with us. My mistake, as I'm reminded every time I see or feel Kit's scar, the valley of shiny flesh with its mountains of scar tissue on either side.

The doorbell rings, and I haul myself to my feet. The postman has a parcel for me to look after most days. Working from home means our house is the porter's lodge for half of Wilbraham

Road. I don't mind, or at least I don't mind now I'm pregnant. And I never minded the bulky stuff; I didn't even mind the garden furniture for number 32 that once sat in my hallway for a whole week. It was the baby stuff that used to pain me, the parcels for Ronni from Mothercare or JoJo Maman Bébé or Petit Bateau. The packages of miniature clothes would mock me, the voice in my head screaming *get them out get them out get them out get them out*.

Our front hall is one of my favourite things about the house. The floor tiles are Minton, all fleur-de-lys and fiddly curlicues – they go for thousands on eBay – and the front door is the original Arts and Crafts, with leaded lights in four panels. I can tell through the coloured glass that it's not the postman but Mac; his profile is quite distinctive these days. He was an early adopter of the now-ubiquitous beard and at the moment he looks rather like D. H. Lawrence, with a huge gingery fuzz that makes Kit's putative beard look like five o'clock shadow.

'To what do I owe the pleasure?' I say, unhooking the chain and throwing the door wide. Mac's wearing hobnail boots, and tweed trousers held up by braces twanged over a short-sleeved shirt. I wouldn't be surprised to find a penny farthing parked behind him. He's carrying a big brown paper bag, the kind Americans put their groceries in on telly, only this one has the Bean/Bone logo printed on it. The forward slash was my idea.

'Decaf latte for the calcium, some sourdough bread, couple of wheatbran muffins for later. And we've been juicing.' He takes out four clear plastic cups with holes for straws in the top: purple, yellow, orange, and a green one that looks like it should have marsh gas coming off it.

'What the fuck is this? Ectoplasm?'

'Hemp and wheatgrass.' He lines them up on the kitchen counter. 'And the *pièce de résistance*.' It's the bone broth that gives his emporium its name and reputation – glorified stock, really, boiled-up bones and carcasses. People round here can't get enough of it.

'I can't look after you as well as Kit does but at least I can feed you up,' says Mac.

'You shouldn't have,' I say, but my mouth waters against my will and I realise I still haven't eaten. 'Are you coming in?'

'I'd better get back,' he says. 'But I'll come around like this with brunch every day while Kit's away, check you're ok. And if you need anything, you just let me know. How are you feeling?'

'I thought I was going to have an anxiety attack this morning just after he went, but I got it under control,' I say.

Mac actually takes a step back. 'Shall I call Ling?' The implication is clear. Anything medical, anything to do with the babies, and he will drop everything and come around. Emotional problems are the women's department. Mac has not mellowed completely.

'No, no,' I say. Ling's a social worker; she'll be knocking on the metal front door to some shitty flat about now, maybe with an interpreter in tow, maybe with the police. I'd have to be desperate to interrupt her during work hours.

'Right, then, I'd better go.' He bends down and gives me a clumsy kiss on the cheek. 'I'm having the girls tonight, so I'll see you again tomorrow.'

Ling and Mac haven't been a couple since we were kids – they broke down, in every sense of the phrase, around the same time everything was happening to us – but they work better apart than they did together, so much so that when Ling got broody again, Mac obliged her with another baby. Their girls, Juno and Piper, live between two homes, both on the Ladder, four streets apart. Three homes if you count our house, where they have their own room, for now at least.

I pour the bone broth down the sink and go back to my tidying. The lamp on our hall table is a light-up globe, an old child's toy that I fell in love with in a charity shop, and I draw the route of Kit's ship, dragging my fingertips through the choppy North Sea. I can cover the whole of the path of totality, where the shadow falls fullest, with my thumb. The Faroe Islands are so tiny that even my little finger obscures them. They look too small to hide in. My arms start to bristle. Beth is a trapdoor; one thought of her and I lose my footing and fall. I pull my sleeves down over my skin and spin the globe until the oceans and the land blur green and blue, and the shadow covers everything.

5

LAURA
10 August 1999

The coach reached its final destination somewhere south of Helston. Local police in fluoro jackets looked us up and down. At the edge of the road, Ling leant against the campervan, head tilted, trying to bask in sun that struggled weakly through gauzy cloud. A hand-painted cardboard sign propped next to her said, Heavy Tent? Lifts To Lizard Point £2.

At the sound of my voice she opened her eyes and broke into a smile.

'I didn't expect chauffeur service,' I said.

'The coaches can only go so far and the site's miles away. Besides, it's a way to claw back some money.'

She held out her hand for coins as people lined up, threw open the van door and the motley crew piled in. At the end of the century, youth tribes were blurring and along with the crusties and the goths were club girls in fairy wings and crisp Essex boys in designer jeans. A dirty red sleeping bag was rolled and strapped in the corner, and the smell of dope was inescapable. Those who couldn't find a seat sat cross-legged on the oily floor, clearly delighted after nine hours in a National Express coach to be able to loll and smoke not only with impunity but encouragement.

I rode shotgun with Ling and put my feet up on the dashboard.

'Is Kit pissed off about the weather?' I asked. She rolled her eyes.

'I've never seen a sulk like it. Me and Mac keep saying, the eclipse will still happen, the festival's still going on, he can either decide to enjoy it or decide not to.'

'He wants it to be perfect,' I said.

'I don't think anything's going to be perfect. Turnout isn't great, because of the weather,' she said. 'Predicted numbers were twenty thousand. Rory – that's the farmer whose land it is – was saying he needs fifteen just to break even and there can't be more than five thousand there. Even allowing for last-minute arrivals, that's pretty shit.'

I sighed. 'Is there any good news?'

Ling wrinkled her nose in thought. 'Well, the chill means that people want hot drinks. We'll still run at a loss, though. We might even pack up a day early, just enjoy the sound systems instead— oh for fuck's sake I've gone past it.'

She slammed on the brakes. I braced my legs but others in the back were thrown hard. 'Sorry!' Ling called over her shoulder. She reversed carefully around a thickly shrubbed corner and then doubled back on herself before turning into an unmade road. 'This is another reason why no one's turned up,' she said, as we bounced over rocks. 'The locals aren't exactly overjoyed about the festival and they've started hiding the signposts. Not just the ones they'd made for Lizard Point Festival but the actual official signposts saying where all the villages are and stuff. I can't tell one dirt track from the next.'

'That's the problem with the countryside,' I said, as we disappeared into a tree tunnel; leafy green shadows swam like fish across the windscreen. 'Not enough proper landmarks. We need a nice McDonald's on the middle of a roundabout or something.'

We came out of the tree tunnel to find a huge perimeter wall of aluminium panels. There were more police, including one on horseback, at the entrance. The van was searched thoroughly for stowaways and casually for drugs by two burly men in high-visibility vests; paper tickets were exchanged for wristbands. Ling and I continued alone, the catering sticker in the van's window giving us access across the fields. The van bumped over uneven ground, past a funfair and a huge blue tent hung with bright gold streamers. All the staples of the music festival were there. Flags, drums, a falafel stall, fairground rides and a bare-chested man on stilts covered in woad. But without a crowd, it just looked like the fall-out of some kind of humanitarian disaster. There was literally tumbleweed wheeling across the yellowed field.

Ling parked the van next to our tents; little red dome for them, bigger green pointy one for me and Kit. I unzipped the door, the familiar sound from camping holidays and festivals past, to see two clean sleeping bags zipped together and laid flat on a double airbed. A drying towel gave off a faint smell of soap.

Our stall was set up under an oak tree, a wide navy tent with an open front. Mac stood next to a bubbling tea urn. A disco ball spinning above him threw diamonds of light across his face and I could smell the soft cinnamon scent of chai, the spiced tea we all drank back then. Wind chimes hanging from the branches tinkled but there were too many of them to be soothing.

'Still time for word to get around,' he said into his mug. But it was less than twenty hours until the shadow.

Kit emerged from the mysterious interior, carrying a rubbish bag. He hadn't shaved since I'd last seen him and his stubble stood out like sparks against skin.

'Hey,' I said softly. He was so locked in gloom that it took a split second for him to take me in; and then a smile transformed him, and I felt the usual pride at being the one to pull him out of a bad mood. He let the bag drop to the floor and, when we kissed, I could feel myself uncoil.

'You smell nicer than I expected,' I said.

'Rory's opened up the farmhouse, you can pay to have a hot shower,' he said.

'Yeah,' scoffed Mac. 'Two days in and the weekender hippies

are discovering the limits of their own hygiene.' He said this as if festering in your own dirt was something to be proud of.

'Ignore him,' said Kit. 'It's the best four quid I ever spent.' He turned to Mac. 'And I don't blame Rory. We're not the only ones losing money this weekend.' He tucked a wisp of hair behind my ears. 'How'd the interview go?'

'Ok, I think. We'll see.'

'I bet you were brilliant,' he said detachedly, looking up as a huge grey cloud scudded overhead.

'The cloud might go,' I said. 'You never know, the weather forecasters get it wrong all the time.' My reassurance bounced off him; he grumbled about clouds and showers until something else caught his attention. 'Oh, what's going on here?' He twiddled a knob on the urn. 'It's buggered again, there's a loose connection round the back. You stay here, have your drink, while I fix this.' He kissed the top of my head and vanished around the back of the tent.

Mac sparked a long, thin joint. I took a long drag to take the edge off London and reset my mind, then passed it to Ling. I could take or leave drugs in those days, I prided myself on it. Addiction already had its fangs sunk deep into Lachlan. I saw it crouching in wait for Mac, and thought myself lucky not to have the disease. I didn't realise of course that my poison was within me, chemicals that my brain could manufacture at the slam of a door or the strike of a match. The stress hormones of adrenaline and cortisol, when pumped in sufficient quantity, rival anything you can smoke or swallow. Within a year of the Lizard, I would envy those who could dry out in rehab. When you suffer from anxiety, you carry an endless supply.

Still, I was pleasantly fuzzy by the time Kit came back, having won his battle with the troublesome hose. Mac waved the joint under his nose.

'Come on, Kit. Snap you out of your strop.'

'I want to keep a clear head for the eclipse,' said Kit haughtily.

'But it's not until tomorrow,' said Ling.

41

'Don't worry,' said Mac, who always took refusal of hospitality in any form personally. 'If he comes as late to drugs as he did to sex and rock 'n' roll, he'll probably have his first E on his fortieth birthday.'

I expected Kit to laugh it off – we'd had our wild nights – but he scowled instead. Only Mac could rile him like this. At some point in their relationship, probably in the ten minutes between Mac's birth and Kit's, probably in their shared womb, it seemed to have been decided that Mac had the balance of power. He had even appropriated their shared surname as his nickname, something no one but me seemed to think was weird. It wasn't that he always had to be right, although more than once I saw Kit dumb down an argument to hasten its end. It was simply that his opinion carried more weight than Kit's.

'I'm going to go and unpack,' I said, and walked across the field, knowing Kit would follow me. We didn't unpack, we went to bed, or rather to sleeping bag. Sex back then was ballast that had to be chucked over the side before we could get on with anything else. Afterwards, we lay in the greeny light of the tent, my knickers in a figure eight round one ankle.

'How far are we from the sea?' I asked.

'About twenty minutes. But if you're in the mood for more of a hike, we could go to Goonhilly Downs. It's where the first ever satellite signal was broadcast. They've got these huge satellite receivers, tall as skyscrapers.'

'It's not *quite* the romantic walk I was hoping for.'

'It is, in its own way,' he said. 'Right in the middle of all this technology, there's a load of standing stones. Megaliths just scattered around. And they built a fucking satellite station there! It's decommissioned now.'

'I love you,' I said. 'But I draw the line at going to see a satellite dish when the Cornish coast is over *there*.'

We flashed our wristbands to escape the festival and took the sea road towards Lizard Point. The tiny town, evidently resting on the laurels of its southernmost location, didn't have much to offer.

It was gridlocked with motorhomes and estate cars; tourists queued for cream teas at the tatty café. A country lane thinned to a craggy footpath. From a distance, the sea was molten lead, then suddenly we were at the cliff edge looking down on aquamarine rockpools.

'You can see why all the smugglers' ships used to get wrecked,' I said as a large wave dragged backwards to reveal jutting black rocks, a dinosaur's jaw.

'I'd make a good smuggler, stolen bounty on the high seas,' Kit said, and we both laughed because it was hard to imagine a less piratical man. 'I could come at you in salty breeches with a cutlass between my teeth.'

'And I could hide my rubies in my petticoats.'

'Oo-ar,' he said, and I had him back. He coiled his hands through my hair and pulled me in close.

'I just want tomorrow to be perfect,' he said.

'There's no such thing as perfect.'

'There's us.'

'Don't be a twat.'

He smiled and let my hair go.

Kit still believes that things went downhill that weekend because of what happened. That if I had turned left instead of right in the moments after the eclipse, we would have continued to sail that perfect golden stream. He is wrong. We were young and we were lucky but we weren't immune to the same shit that happens to everyone else. Even – especially – good sex is unsustainable. Time and the mundanities of living would have stripped the gloss eventually. If anything, we are so strong now *because* of the trauma that forged us. But Kit won't be persuaded. Despite that theory of parallel lives he's always on about, where there are infinite universes where all possible actions take place, you cannot live the same life twice, going back and redoing things differently. We'll never know what our relationship might have been like, untested. We only have the one we have.

6

KIT
18 March 2015

England blends grey to green as my train rattles north out of London, and I get calmer with every mile. It's actually a physiological process; I'm aware of my vertebrae unknotting themselves one by one. At first I put it down to the relief of hitting every mark on my schedule, unimpeded by failed signals or passenger action, but as I unwrap the sandwich that Laura made for me, and the thought of her brings some of the tightness back to my chest, I realise it goes deeper than that. It is, I understand with a horrible jolt, relief at being *away* from my wife. Four days away from the mood swings and the paranoia and the endless grisly speculation. Four days when I only have to look out for myself.

Thinking about Laura in these terms takes away my appetite. She can't help her anxiety. I know that it's torture for her. It kills me to see her clawing and crying, watching worry eat her up. I've asked her if she can't talk to a counsellor again – we'd find the money; see if there's a way she can learn to file the past away, like you might an old document that you may never use, but you need to keep just in case. Laura won't countenance it. She doesn't have that kind of brain, as she keeps telling me. And when I see her, scratching away at her arms, or breathing carefully in time to some internal mantra, I'm grateful I don't have her kind of imagination. I should be grateful that one of us is able to keep it together. Guilt crashes in on the wake of these feelings, a futile emotion, and I force my mind elsewhere.

The train rips through Nottinghamshire, Beth's home county. Pylons string electric cables across gentle hills. I've always thought that pylons are the ultimate measure of how adaptable human beings are. We've got these gigantic steel monsters marching through our countryside and not only do we not run screaming, we don't even see them any more.

We stop for no apparent reason near Newark. The train's silence lays bare a looping whisper in my head: *You shouldn't go.* My conscience has my wife's voice and mannerisms, as well as most of her convictions.

I check my phone. Nothing from Laura, which I hope means she's gone back to bed. I roll through the screens. Yesterday I added three pages of new icons, shortcuts to every eclipse-chasing blog, chatroom and forum I could find online. I want to be able to measure the official weather reports against the rumours.

Then, checking over my shoulder out of force of habit, I dip into the secret Facebook page I've hidden in the 'utilities' tile, behind a dozen other apps I'll never use. Laura would *kill* me if she knew about this, but there's no better place than Facebook to find out what's happening, and I've made myself as anonymous as possible; made-up name, an avatar that doesn't show my face and all the location services turned off. I only access it on my phone or my PC in the study, never on the shared tablet. There was a near miss a couple of years ago when a woman calling herself ShadyLady (I guess two can play the fake name game), her picture a shapely silhouette before a crescent sun, sent me a private message saying, Are you Kit McCall? I blocked her, then deactivated the account for twelve months, and she's never bothered me since I came back.

On the group's wall, the mood ranges from cautious optimism to unbridled woe. New worries cloud out the old, and by the time my train pulls into Newcastle Station, my thoughts are only of the sky.

'Chris!' There's a satellite delay of half a second, as there always is when anyone uses my public name.

'Richard!' He's underneath the clock, resplendent in his Faroese jumper. His backpack is smaller than mine and he's carrying a little crate of Newcastle Brown Ale. He waves it in salute when he sees me, and we shake hands; it isn't a hugging sort of friendship, although that might change after four days sharing a berth. He worked in my lab a few years ago, and after discovering a shared affinity for cult fiction, we went out for the odd drink after work. He's more of a telescope astronomer than an eclipse chaser, but when it became clear that Laura couldn't travel to Tórshavn, I asked Richard, not just for the company but to cover her half of the fare; we are counting every penny now. Richard doesn't know I was ever Kit McCall and he certainly doesn't know about Beth. Laura wondered once if he would be in any danger as my travelling companion, but why would he?

'It's bright orange,' he says, gazing in wonderment at my beard. We fall into step on the way to the pick-up point, where a minibus waits with its motor running.

In our seats, the conversation takes a meteorological turn, and as we move on to the nerdy intricacies of thermal fronts and planetary alignment, I feel like I'm slowly being lowered into a warm bath. There's no need for either of us to stop and explain a theory or phenomenon. Laura's got the eclipse-chasing bug, but there's only so much science she can absorb before her eyes start glazing over. For her, it's enough to stand and observe in awe. I'll never understand that, although I've learned to respect it. But the way I feel now, discussing celestial mechanics with someone on my level, I can only liken to how it must be like living in a country where no one else speaks your mother tongue. You can get by in the foreign language, you can communicate, but the pleasure of speaking to those who understand every subtlety and nuance must make you want to cry with relief. There's an objectivity too in these debates. Even in smooth marriages, conversation is never neutral; everything you say carries the weight of every conversation you've ever had. Pure science is respite precisely because there's no inherent morality in fact. Everything I say to Laura is picked up and examined for an ethical content that baffles me more often than not. You're on safe ground with knowledge, with quantifiable data. Opinions on the other hand, are shifting, baseless things. Sometimes I think I don't have a true opinion about anything except Laura.

'Got an *awesome* geolocation tracking app on my tablet,' says Richard, showing me a screen with statistics superimposed over a world map. 'Total, partial, annular, it's all there.' I suppress a stab of irritation; *I'm* supposed to be introducing *him* to the experience.

'You boys know your stuff,' says a middle-aged man seated in front of us. 'Ever done anything like this before?'

I can't help but puff out my chest. 'This is my twelfth total eclipse.'

All eyebrows rise in unison and I feel like a god among mortals.

'I'm an eclipse virgin,' says Richard cheerfully.

'Us too,' says the man, gesturing to his wife.

'It's been a long time since I was a virgin,' she says, to cackles all round.

I call Mac and check he doesn't plan to stray too far from our neighbourhood while I'm away. If there's an emergency I want to make sure it's Mac who takes her to hospital. I want her to be with family. Ling can't always control when she works, my mum would only flap, and Laura's dad won't be able to get to her in time.

'I promise,' he says. 'I won't leave the Ladder till you come home. Don't worry, I'll look after her.'

'Stay driving sober,' I say. It's a joke, of course. He's been dry for fourteen years now. It wasn't sobriety or even fatherhood that mellowed him but success. Laura and I concluded that it was keeping his inner capitalist suffocated under a duvet of hippy liberalism that was making him so bitter. He calls himself a bare-foot entrepreneur, which makes him sound like a dick, and the difference in me is that these days I actually tell him that.

The cityscape changes as we approach the docks. Shipping crates are piled high as tower blocks and cranes like rocket launch pads pierce the low cloud. I recognise the *Princess Celeste* from the brochure and feel the squeeze of claustrophobia when I see how tiny the windows are.

At the harbourside, our luggage is whisked away, with reassurances that it will be waiting for us in our cabins. We stand, Richard and I, in matching knitwear, at the foot of the gangplank. There are hundreds of people around us. Despite my reassurances to Laura, I find myself on the lookout for dark hair that curls like smoke, and my ears are braced for the sound of my old name.

7

LAURA

11 August 1999

The day of the eclipse dawned dull and chilly. We woke up at eight, even though we'd been working till midnight and after that we'd gone dancing. A girl with a pot of golden body paint gave me and Ling a two-for-one deal: we had had flaming suns painted on our bare arms even though it was so cold that our bodies gave off steam. We'd found a little tent playing trance and gone wild. Now most of the body paint was on my hands and the sleeping bag; still, that smudged golden sun was the only one we were likely to see.

When Kit poked his head out of the tent I thought he was going to cry. 'I've never been clouded out before,' he said. 'I know it happens, there's a one in six chance, but I just can't see how it'll be the same.'

An hour before first contact, I packed a little bag and Kit checked his camera for the millionth time. We wandered past the Waltzer and the Ferris wheel, through the trees to the stall. In the absence of customers, Mac and Ling were in the chill-out area, giggling hysterically. 'Yo!' said Mac in greeting. I studied them like a vice-squad cop; their eyes were pin-sharp, so not dope: their jaws were still, so not E; so acid, which meant they were good for nothing for the rest of the day.

'You're taking the *piss*,' said Kit. I knew it wasn't prudishness, or even frustration at the loss of money, so much as anger at Mac's lack of respect for the phenomenon. 'Let's leave them to it,' he said to me. 'I couldn't care less about making a profit now.'

They didn't even notice us go.

The main stage was as busy as we'd seen it, the field packed full of people nodding in time to thin trance music and squinting hopefully at the white sky. Many of them were wearing their protective goggles, Mylar-coated lenses in cardboard frames, even though there would be nothing to see for some time. Occasional shafts of light broke through, to sparse whoops and whistles that died away as the clouds closed over. Kit looked nervously around the crowd.

'There's no horizon here,' he said. 'If we're going to be clouded out, then we want to be able to see as much sky as possible.'

We turned slowly in a circle.

'What's on the other side of those trees?' I asked. 'There might be a better view there.'

The other side of those trees turned out to be full of parked vans and the HGVs that had brought the funfair equipment. There was an abandoned dodgem whose seats had been slashed so that the stuffing was foaming out; a vital-looking piece of equipment, something that looked like the whole arm from one of those spider rides. I resolved not to go on a single ride. Behind all these was the perimeter fence, cutting off the sky at twenty feet.

'It's worse here than at the stage,' grumbled Kit.

'Hang on,' I said. There was a lorry parked right next to the fence, its roof level with the top of it. I looked at Kit, then up at the top.

'We *can't*,' he said, but he did a recce of the vehicle, looking first in the driver's cab and then in through the windows before giving me the thumbs-up. He was up in one graceful bound; I clambered like a monkey, my fingers clinging to the wing mirrors, and my feet finding purchase on the bottom of the windscreen, until Kit pulled me up the last few feet.

Even on an overcast day, the view was a picture painted just for us. Green hills rolled down to the sea in the distance. Where yesterday we had had the clifftops to ourselves, today the grass and heather were stubbled with tourists. Through some trick of the air, the music sounded better up here than it did by the stage, the doof-doof bass less fuzzy, the electric treble cleaner. I took the eclipse goggles from my jeans pocket and wiped the plasticky lenses on the hem of my jumper. One of our customers last night had told us there was a shortage of the glasses on site; apparently pairs were changing hands for up to fifty pounds. I took mine off; they were just a strip of paper and plastic in my hand. I wondered how often an object is priceless one minute and valueless the next.

Kit's melancholy had given way to agitation now. He held my hand so tight that I had to pull it away.

'Sorry,' he said, rubbing my crushed knuckles back to life.

Then the winds began.

Kit had of course told me about the eclipse winds, an unearthly portent that can range from a breeze to a near-hurricane. It whipped my hair into silver streamers that Kit smoothed down with his hands then caught and held at the nape of my neck. It was storybook weather, a prelude to a fairy tale. 'It's coming,' he said. Without sight of the sun, the leaching of the light was slow and undramatic; dusk, remarkable only for its eerie timing. Behind us, the festival continued, a screeching treble and a dirty bass building to a crescendo the experience didn't seem to warrant. Every now and then someone would shout 'Come on sun!' as if they were cheering on a Brit in the Wimbledon final. Despite the slicing wind, the clouds above remained a solid mass.

'There.' Kit nodded to his left, and pointed his camera. I followed his gaze and lost my breath. A wall of night pressed in towards us from the Atlantic, a black veil being dragged across the sky. I gasped like I was falling. A lone starling in a tree began a frantic chirping, and the music reached a screeching climax, where I had expected a reverential hush. (I would have the opposite feeling a few years later when we travelled to Tromsø to see the

Northern Lights; I had been surprised at their silence, that they didn't make a whistling noise or crack like whips as they sliced through the air.) Somewhere, far inland, fireworks sounded.

'I didn't know the darkness could be so beautiful,' said Kit, aiming his lens at the horizon.

As if he had summoned it, at that moment, a hole was torn lengthways through the cloud and the sun was partly visible, a sooty black disc surrounded by a ring of pure light. Kit's camera clicked and reloaded next to my ear. An ecstatic cheer carried on the strange winds from all around us. There were none of the phenomena I'd hoped for: no shooting corona, no sun leaking through the moon's craters to create the diamond ring effect, and in a few seconds it was gone, but still I felt changed, as if a giant hand had reached down from the sky and touched me. I was torn between wanting it to be over so that we could talk about it and never wanting it to end. But it did end; the veil pushed east and the colours came back.

I felt suddenly *shy* around Kit, after the weird heavenly intimacy of what we'd just experienced.

'I don't know what to do with myself now,' I said.

He screwed the lens cap back on tightly.

'I've got a *massive* boner,' he suggested.

I laughed, and let him guide me down from the top of the lorry, where I landed in his arms with a thud that toppled him over. We were wrapped together so tightly that the only way to walk was to fall into step, as though in a three-legged race. I had to watch where I was treading; if I hadn't, I might never have seen the purse. It was a little zipper wallet made of brightly coloured wool in an Aztec pattern. I bent to pick it up; there were three five-pound notes inside, and some coppers, but no ID.

'Maybe leave it there in case they come back,' said Kit.

'But *anyone* could pick it up. That could be all someone's money. All they've got for the rest of the festival. All they've got to get home. There's that police Portakabin thing near the entrance. We should give it in there, if we don't see anyone on the way.'

'All right, if it makes you happy.' Kit rolled his eyes. 'I'll go that way and see if I can see anyone who looks like they've just lost a purse.'

'Thanks,' I said, distracted; I'd noticed a coin on the ground a few yards away, and then another.

We dropped hands, and it was the last time everything was perfect.

I have replayed that moment in my head so many times since then. If I could live the Lizard again, would I pick up the purse? There is part of me – the cocksure insistence of hindsight – that says I should have left it on the ground and gone back with Kit. But even knowing what followed, I don't think I could have walked on by. Perhaps, though, I would have gripped Kit a little tighter, for a heartbeat longer, and savoured perfection while I held it in my hand.

8

LAURA
18 March 2015

Insomniacs know that when you wake up too early, everyone else's breakfast time feels like your lunchtime. Bored, I bring my usual mid-afternoon call to my dad forward by a few hours.

I didn't really expect him to pick up. He'll be making the most of the morning rush-hour, standing on a street corner, vainly trying to press leaflets into commuters' unyielding fists.

I text Ling.

Can you talk?

She fires back:

I'm in a case meeting.

And, with that, I've exhausted the list of people I can call for a casual chat. It's not that I worry how few friends I have, just that every now and then I notice it. The babies will change all that. Ronni next door once told me that children were an even better social lubricant than wine.

Suddenly, I realise what's wrong. I've been so preoccupied with Kit's trip that I haven't said good morning to my mother. I pick up the black and white photograph in its tatty wooden frame and kiss the glass.

In March 1982, thirty thousand women joined hands around the perimeter fence of an RAF base at Greenham Common, Berkshire, in protest against nuclear weaponry. I was one of them.

In fact I was in the local newspaper, a smudged newsprint smile on top of patched dungarees. *We shall overcome: four-year-old Laura Langrishe pictured at the Greenham Common Women's Peace Camp with her mother, Wendy.* I keep a framed copy of it on my desk; my father still has the yellowing original on his. Next to it, there's another photograph taken later that week, a white-bordered snapshot, not the original this time – that went the same way as Kit's first map – but a reprint from a negative. In the photograph, I'm outside a tent, wrapped in my mother's skinny arms. She's wearing a paisley headscarf and hooped earrings and there's a hand-rolled cigarette tucked behind her left ear. We are both laughing, matching dimples high on our right cheeks. She was killed by a drunk-driver four weeks later, three stripes deep on a zebra crossing, on the way to pick me up from nursery school.

My dad, Steve, talked, still talks, about Wendy all the time; death pickled her in perfection, so my early years seem idyllic. I would like to remember her shortcomings, but that's never been an option. This is something I didn't realise until embarrassingly recently. I've asked Dad what they rowed about and he says, nothing really, we had the same outlook on life. Maybe it *was* perfect, in those early days. Maybe time would have had her grounding me, policing my wardrobe and friends, disapproving of the music I listened to and the books I read or failed to read. I do know that Wendy carried the newborn me everywhere in a sling decades before this was fashionable, and that I learned to speak by naming the wild flowers on our nature walks. Dad talks fondly of the biscuit-baking and potato-printing we did together at the kitchen table in our rented Croydon flat. I wish I could remember just one of these occasions and trust it as a true memory rather than an internalised legend, but there are just associations and triggers. A witch-cackle laugh; the smell of rolling tobacco and Timotei shampoo. There's only one memory I'm sure is mine, and that's of her doing my hair, whispering that it was too lovely to cut, of brushing it and plaiting it, and I remember purring like a kitten. I certainly didn't get this memory from Dad, who tortured me with years of tangled ponytails yanked high and lopsided in rubber bands. I never asked him if this was true in case it turns out not to be, but I shared the memory with Kit our second evening together, through noisy helpless tears. That night, he brushed out my hair one hundred times before bed.

I might not have learned the practicalities of motherhood from Wendy but there is something of her in me; I feel it, behind my ribs. I like to think of it as all the love she never got to give me. A vacuum should be weightless but I stagger under the burden of it, and I won't feel complete until I pass it on. Adele, with kind but clumsy intentions, says she has enough love for two grandmothers. She certainly has enough wool. Her job suits her, a rag-bag of fabric and yarns; there's nothing to Adele, no opinions on anything outside the home. She and Wendy would have been perfect sitcom warring mothers-in-law.

Wendy was a card-carrying, almost clichéd feminist: as well as camping out at Greenham, she subscribed to *Spare Rib*, and didn't wear a bra, although like me, or like me until I had to start messing around with my hormones, she had the figure of a ten-year-old boy and didn't need to. My dad is more of an accidental feminist. He brought me up with no concessions to gender. He was too busy playing politics to buy me Barbies. He worked as a sub-editor on local papers, then at Fleet Street, and was big in the Trades Union movement throughout my childhood. One early memory I *can* trust is the terrifying crush at the picket line at Wapping when Dad and his friends came out in support of the striking printers. He got quite a lot of things wrong: I spent my ninth birthday at the Trades Union Congress annual conference, for example. But he got a lot right, too. My parents met playing pool in their polytechnic and when I was ten I begged Dad to teach me how to play. He took me for a Coke and a packet of salt and shake crisps at the Croydon Working Men's Club. At first, I stood on an upturned Britvic crate to reach the table. By my thirteenth birthday I was as tall as Dad, and could clear the baize in five minutes.

Since he retired, he's got seriously into party politics. Just before I met Kit, he said, 'The South is dead,' and declared that the only hope for the Left was in the north. He swapped the Croydon flat for a two-up two-down in Toxteth, Liverpool, where he campaigns for the TUSC, a hardline party of the working people. His unelectability in a world ruled by the markets only makes me love him all the more. He should be able to get a good view of the eclipse on Friday. The farther north you go, the greater the shadow. In Toxteth there will be about 90 per cent coverage. Although, as most eclipse chasers will tell you, nothing less than totality counts. Partial eclipses are interesting but they don't give you the shivers. Even 98 per cent coverage is like being *nearly* pregnant.

We used to go whole weeks without speaking, but Dad rings me most days now, ostensibly for help with a clue in the *Guardian* Quick Crossword. Kit thinks he completes the whole crossword in five minutes, then singles out the clue that took him the longest to solve and uses it as a pretext to ring me. Kit finds this endearing; I do, too. Dad loves his daily update on the babies, even if it's just a 'good'. He melts like butter at the mention of them.

He knows about Jamie's trial and he knows what happened later but not that the two are connected.

He doesn't know that the night we escaped with our lives was anything other than a random accident, and he certainly doesn't know that we've been living in fear of a second, successful attack ever since.

He doesn't know that Beth is the reason behind our name change.

Kit and I didn't sit down and decide, as such, that we would keep Beth's attempt to destroy us to ourselves. Rather, it was understood. We talked around the decision. We told each other that we didn't want our families to worry, and it's true that the McCalls already had plenty on their plates. I can see now that my dad's tragedy was years in the past, and he would have been there for me. Sometimes I think I could have told him part of it, but that's the thing with secrets. They're leaky; you can't decide to share the bits that suit you without a million questions oozing out. You have to solder a part of yourself shut.

To understand why Beth is so enraged, Dad would need to understand the context, and all roads lead back to my lie. I put myself on the line for her. I know from my own counselling that anger always has hurt at its core. Until the night I challenged Beth, I was the only one who had never said the wrong thing. Doubt coming from me was worse than from a stranger.

As far as she was concerned, the person she trusted most in the world first betrayed her and then vanished. In saving ourselves, Kit and I took away Beth's right to reply. My psychotherapist would doubtless say we took away her closure.

It seemed like the only option at the time, but rather than dousing her fire it has stoked it.

That I was right doesn't seem to come into it. By then, Beth's sense of right and wrong had been beaten out of shape. Even mine, when I examine it, now seems unrecognisable from the inflexible moral code my father instilled in me, and which he, and my husband, still think I have always lived by.

9

LAURA
11 August 1999

Kit walked one way, while I followed a trail of copper coins towards a cluster of shut-up caravans. An old carousel horse was propped against the nearest one, as though it had galloped away from its rotating prison and finally exhausted itself. The name Eloise was hand-painted on to a scroll on its flank. Just beyond, there was a shuffling noise and a flicker of movement.

'Hi, is this yours?' I said, then let the purse drop to my feet.

The woman was lying face down, her clothes – a long skirt, at first glance – pushed aside. The man was on top. Nothing unusual about that, Kit and I did it all the time. But the expressions on their frozen faces were a million miles from anything I could identify with. The man's back was a cobra's arch. His unfocused eyes were narrowed to slits and spittle hung from lips curled in a monster's snarl, darker than anything I recognised as desire. It made sudden, sickening sense in context of the girl's face. She was looking straight at me; wild eyes locked on to mine. Crude eye make-up streaked her cheeks. Her fingers clawed the earth. That kind of animal terror is something you recognise when you see it, you don't need to have experienced it. Snot poured from one nostril; mud and fragments of leaf and twig were smeared into the mucus, as though her face had been not just brushed against the ground but pushed into it. I knew what I was seeing. The word was loud and ugly in my head. Four capital letters daubed red on a wall, too big to read, too frightening to say.

'Oh Christ,' I said. People talk about blood running cold but mine flushed hot then, it scorched my veins. 'Are you ok?' It sounded pathetic.

At this, the man's head snapped up, and for a moment that terrifying expression was directed at me. I stumbled backwards and gasped to feel the ridged steel of a caravan wall, cold against my back. I don't know how long we stayed locked in that tableau. It could have been thirty seconds; it could have been three. I do know that the gap between what I was seeing and my ability to deal with it seemed blown wide, as though detonated.

'God, sorry,' he said. 'Embarrassing! She's fine. Aren't you?'

The girl blinked at me, but made no move to speak, or even to wipe her face. He pushed himself up and off her. A tensile milky thread trailed from the tip of his penis to her buttocks, then snapped as he tucked a withering erection into his fly with a wince. He stood up from kneeling. Everything about him, from his anorak to his jeans, looked box-fresh, forcedly casual. Brand names in large letters ran across his chest and down his arms. His light brown hair was teased into tidy little peaks. Only muddy patches on his knees and the heels of his hands gave him away.

'This is awkward,' he said, with a nervous laugh. When he smiled, I was horrified to realise that he was beautiful.

The girl lay motionless on the floor. Her left leg and buttock were exposed. At first I thought her skirt had been torn, but the fabric wasn't frayed; then I saw that she was wearing Thai fisherman's trousers, long wraparound pants that we all wore back then. Laid out flat, they looked like a tesseract with ribbons attached; they were an alternative fashion shibboleth, an unfathomable puzzle for the uninitiated that made wearing a sari look as simple as a T-shirt, but once you knew how to put them on, it was easy. They had no Lycra or give in them; they must have been yanked, very hard, to one side to reveal that much flesh.

I looked over my shoulder for Kit but he wasn't there.

'Are you hurt?' I asked her. 'Has he hurt you?' She blinked at me and I wondered if she was on something. 'What've you done?' I said to the man.

'You've got the wrong end of the stick,' he told me, though he didn't offer me an alternative. He turned to the girl and said kindly, cajolingly, 'Come *on*, baby.'

I tiptoed closer to the girl and put out my hand to help her up but she didn't move. 'What's your name?' She shrank back and crouched against the wheel of one of the caravans. I thought about covering her with my cardigan but it would probably be covered in my hair and Kit's and could only contaminate a crime scene; even without naming the crime I was already thinking in terms of forensics. 'It's ok,' I said, feeling hopelessly out of my depth.

'Laura?' Kit's voice was clear and close. 'I couldn't find anything.' The caravan walls made a narrow corridor; the man walked backwards along it, away from me, away from his victim and backed straight into Kit.

'Whoa!' said Kit. 'You want to look where you're—'

The man's yelp, coupled no doubt with the expression on my face, made Kit realise how serious the situation was. 'What's happened? Are you all right?'

I took a step forward and put myself between the men and the girl. 'There's a girl, I think she's been . . .' The word fell apart on my lips, the letters in a domino topple. 'I think she's been attacked.'

The man rolled his eyes in my direction. 'It's fine.' He threw Kit a smile of all-boys-together conspiracy. 'We just weren't expecting company. Were we?' The girl wiped her nose on the back of her hand, looking without expression at the smear of mucus on her sleeve. 'She's just embarrassed to be caught with her knickers down, aren't you?' His voice was light but his jaw flexed between sentences. 'I'm not exactly delighted about it myself. But that's all there is to it. Your missus has jumped to the wrong conclusion.'

'Oh,' said Kit uncertainly.

'I know what I saw,' I said.

The man began to back away. With the girl mute and Kit confused, I realised it was up to me.

'I think we need to get the police involved.' It came out firmly, my voice betraying none of the roaring terror inside.

'*You* need to calm down,' said the man, but he was losing his own cool.

I stood my ground. 'If you haven't done anything wrong, you haven't got anything to worry about.'

He rounded on the girl. 'Would you fucking *say* something so we can all get on with our lives?' There was violence in his voice. To me, it was as good as a confession and Kit's face at last registered the seriousness of what was happening. The man realised he'd lost his ally.

'Fuck this.' He walked briskly away, past the fairground junk and into the trees.

'Kit, don't let him get away!' I said. 'Go after him!'

'What?' He looked absolutely horrified but he did it. My shy, gentle Kit ran after a violent man because I asked him to and because he took me at my word that something terrible had happened.

I crouched next to the girl. 'Oh, you poor thing,' I said. 'Don't worry, we'll make it all right.' My arm brushed against hers; soft skin over softer flesh. I got a proper look at her; green irises were almost eclipsed by flared pupils. Inside the cloud of black hair was a heart-shaped face, pinched but pretty. She looked a bit like Disney's Snow White if she had grown out her bob and loosened her corset.

'Who is he?' I asked her. 'D'you know him?'

She opened her mouth to speak but only a strangled croak came out. Long black hair was all over her clothes. She picked up a strand, put her hand up to one of her temples, then took her fingers away like she was expecting blood.

'Did he pull your hair out?' She didn't answer, just let the strands float down to the ground. 'Christ. Ok. We need to go and find some police,' I said. 'There's a little Portakabin at the gate. Can you get there?'

This time she managed to shake her head.

'Can you tell me your name?'

'Beth.' She nodded, as though glad she'd remembered it.

'All right Beth, I'm going to call them for you.' My phone was in my bag, but I'd turned it off on arrival. I held down the power button and waited for the little dot matrix screen to glow green. It took so long that I was resigned to waiting for Kit to return then sending him off to get help, or even hoping that he would somehow manage to get the man to the police on his own, but I couldn't see that happening. For the first time I felt a chill of fear on Kit's behalf. Would he *hit* Kit? At last, the phone lit up. I pressed the rubbery nine key three times, but nothing happened. I checked the screen; the bars were all empty. I waved it around in the air to summon a signal.

'I need to step just a few feet away, so that I can get reception,' I said. 'I won't go far.' I had to walk a good twenty paces, to the broken bumper car, before the connection was made; all I could see of Beth was a silver Chipie trainer poking from the doorway of the caravan.

'999, what's your emergency?' The voice was female: West Country; young. There was all the usual office hubbub in the background, and it was strange to imagine that this was all in a day's work for the woman on the other end of the line; that she might be drinking tea while I talked.

'I'm at the Lizard Point festival, near Helston, and I need to report . . .' I choked on the word, then gathered myself. 'I need to report a rape. Not me, not me. I found a girl, and he was . . . we need the police.'

I picked at the foam stuffing spilling out of the bumper car's torn seat.

'Is the victim conscious?'

'God yeah, she can walk, she hasn't got any cuts or anything, but I think she's . . . she's traumatised, she's not really talking properly. I'd say she needs an ambulance. Can it be a WPC? Can it be a female paramedic?'

I craned my neck through the trees, to get a sense of what might be happening on the other side. I thought I could hear the murmur of a moving crowd, but no individual voices.

'We'll send the first available officers,' said the operator. 'Just stay with her.'

'The man who did it, I saw him, but he's run off.' The wad of foam in my hand looked like candy floss. I let it float away.

'Would you be able to give a description?'

'She's called Beth; she's got black hair and—'

'Of the attacker.'

'Oh. Yes.' I could still see him with photographic clarity. 'He's got short brown hair in spikes, navy Diesel jacket, Levi's twist jeans, white Adidas shell-toes.' It was surreal, ridiculous, like I was reading aloud a fashion spread from *Loaded* magazine. A sweaty hand on my arm nearly made me drop the phone. It was Kit, his breath ragged.

'Hang on,' I said to the operator, and covered the receiver while Kit mouthed words at me.

'He just disappeared into the crowd,' he said.

I repeated, then confirmed our location using a tall Water Aid flag as a marker, then ended the call.

'He was halfway across the field before I even set off,' said Kit. 'I'm so sorry.'

'Hey,' I said. 'You did your best.'

'Shit.' He nodded towards the caravan. 'Is she ok? Has she said anything yet?'

I shook my head. 'Do you think they'll catch him?'

'I don't know. I can't . . .' he spread his hands and looked down at empty palms as though the answer might be there, then shrugged helplessly. 'I haven't got a clue.'

'I'd better tell her they're on their way.'

Beth nodded at the news and muttered, 'Thanks.' I sat next to her on the cold steel step of the caravan and waited for ten minutes that felt like an eternity.

'It is mine,' she said suddenly.

'I'm sorry?'

'That purse. It's mine. It must have fallen out of my pocket.'

I looked down, faintly surprised to see it still in my hand.

'Here.' I had to close her fingers around it.

At last, Kit called out that the police were on their way. There were two of them, a stocky man with a shaved head and a reedy woman with permed mousy hair.

'She's here,' I said, rather unnecessarily.

'All right, love,' said the female PC, crouching to our level. 'It's Beth, isn't it? Is that right?' Beth nodded. 'Ok, don't worry Beth. We're going to make you safe. We're going to take you to a suite at the police station, where we'll get a doctor to look you over.'

'Will it be a female doctor?' I said.

'We've requested one,' said the PC, but she was frowning. 'Beth, do you need someone to accompany you?' She looked at me, but Beth was already shaking her head.

'We don't know each other,' I said. 'I just stumbled across it all.'

The male officer cleared his throat. 'We'll take it from here,' he told me. 'I've got a colleague on the way to take a statement from both of you.'

Kit and I sat on the bonnet of the bumper car waiting to be interviewed. I picked harder at the broken seats while he twiddled a blade of grass between his fingers.

'Why would he *do* that, just after an eclipse?'

I looked at him, appalled. 'Or at any time. Jesus Christ, Kit. I can't believe you said that.'

Things sometimes didn't reach deep into Kit the way they seemed to with me, although once you pointed out why you were upset or angry, he always saw your point. He was, to his credit, mortified. 'Yeah – sorry – I didn't mean it like that . . . I just don't get why—'

'I know,' I said, blowing out my cheeks. 'We're both in shock, I suppose.' I tried to balance a little cloud of foam on my palm but my hands were shaking too much. 'I should've got there sooner.'

'Maybe you stopped him before he did anything worse,' said Kit.

I let that thought sink in.

'Are you my witnesses?' Standing before us was a woman who looked like she'd just stepped from the streets of 1980s London. She wore a shiny black suit and had cropped blonde hair that meant business, harsh bright make-up that had been in fashion when she was about twenty and she was fucked if she had time to get a makeover.

'DS Carol Kent, Devon and Cornwall Police,' she said in a voice that had us both on our feet. 'If you could accompany me back to the festival station, I can take your statements.'

There were two scuffed Portakabins side by side near the main stage. In mine, there was a police dog, a beautiful Alsatian, sitting with his handler. He grew excited, sniffing the air around me and straining on his lead, and my cheeks glowed as I realised he was probably picking up some trace of yesterday's smoke. Someone gave me a cup of weak tea and I told DS Kent exactly what I'd seen, starting with the purse and including as many details as I could, although I grew clumsy and inarticulate when it came to describing the expression on the man's face. They asked me the same questions over and over, as though testing me for consistency. I kept repeating variations of the same phrase. 'If you could have seen him . . . if you'd *been* there, you'd understand.'

In the lull while my words were transcribed I heard snatches of Kit giving his statement next door. I caught him saying that he hadn't seen the attack, only its aftermath, and I knew that he had gone into scientist mode. Observe, record, without prejudice. At that moment, I wished with all my heart that he had seen what I had seen behind the caravans, although later, when madness took me over, I was to be glad that he hadn't.

Our statements were written down, read back to us; after we had given our home addresses, we were free to go.

'Where's Beth?' I asked. 'What's going to happen to her?'

'She's been made safe,' said Kent with an air of finality.

It was late afternoon when we finally signed our statements. Even though the festival had another night left, people were already taking down their tents and loading up their roof racks. Rumours blew through the crowd. 'Someone's OD'd in a field round the back.' 'No, apparently it was a mugging.' 'I heard it was a fight.' No one came close to the truth.

'I could do with a drink,' said Kit. The main bar wasn't licensed to serve spirits so we made do with the strongest thing we could find, a local gut-rot cider so potent they only sold it by the half-pint. We drank two each, too quickly, sitting cross-legged on the grass. Neither of us said it, but we were both watching the crowd for Beth's attacker.

Even after two days of relentless dance music, the pop of the police siren made everyone jump. The crowd parted to let a patrol car through. Coming from the direction of the police cabins, it rumbled along at walking speed. Everyone was gawping through the back windows but Kit and I alone must have known the significance – if not yet the identity – of the passenger. He was leaning away from the glass, but I could still see him in memorised profile, spiky brown hair on top of a navy blue jacket. If I hadn't already been sitting down I would have collapsed with relief.

'They've got him,' I said.

10

KIT
18 March 2015

'This is cosy,' I say, as the horizon tilts nauseatingly through a porthole the size of a compact disc. The cabin is tiny. An hour at sea and I'm already feeling the claustrophobia of the watertight compartment. Richard whistles through his nose when he breathes. I feel pathetically, childishly, homesick for my wife. I even miss that third wheel in our relationship, her anxiety.

Richard is poring over the brochure. 'There's a *casino* on board,' he says in wonderment, as though the *Princess Celeste* was not a working cruise ship but a vessel commissioned especially for astronomers, unconcerned with such earthly pursuits, and he was expecting a state-of-the-art observatory next to the games room. There is also a ballroom, which will host tonight's disco (goodness knows what that'll be like), a miniature cinema and a beauty parlour. Before departure, I couldn't see myself setting foot in any of them. Now I realise that the alternative is sitting in this cramped room, even the quoits deck has a certain allure.

'Gone are the days when I could uncap one of these with my teeth,' says Richard, twisting away at his bottle of Newcastle Brown Ale with his bare hands. 'I'm going to take the skin off my palms if I'm not careful.' He realises what he's said and his eyes travel to my hands, which are folded on my lap. 'Oh, Chris, sorry, I didn't think.'

'It's fine,' I say, because he's more embarrassed than I am. 'There's a Swiss Army in the side pocket of my rucksack – no, the other one – that's it.' There's a comforting plink as Richard opens a beer for each of us. He folds the bottle opener back into the body of the knife, and when he throws it to me I catch it in one hand, then rise to take the proffered bottle.

'Let's take these out for a walk,' I say.

Outside, our fellow passengers are determinedly taking their afternoon promenade, hoods up against a biting salty wind. Even with their hair and faces hidden, I can tell by their gaits that most of them are retirement age. This is the first eclipse package tour I've ever been on – for obvious reasons, Laura and I always travelled independently – and I feel the experienced traveller's instinctive disdain for the mere tourist. As soon as I notice this attitude I vow to overcome it on this trip. An eclipse viewed from the deck of a ship or outside a coach is no less valid than one witnessed in a field full of hippies after a dusty Jeep-ride across the desert, or one viewed alone on top of a mountain. I'm nearly forty and about to become a father. I congratulate myself on my own maturity. It occurs to me for the first time, and I can't believe this realisation has taken so long, that one good thing about being so removed from the alternative scene is that those few people who recognise us – and as discreet as we tried to be, there was at least one occasion when we could not avoid drawing attention to ourselves – are less likely to be here. That old granny there in her orthopaedic shoes, for example, I'll stake my (re)mortgage that she won't have witnessed the scene in Zambia.

A public address system bing-bongs into life, and everyone freezes mid-step as a fruity, weirdly familiar male voice broadcasts a message about tonight's introductory drinks, lecture and – 'for those who still have the stamina' – disco. My heart bolts. If Beth is here, this is surely where she'll reveal herself. I know from Zambia that she doesn't mind making a scene. The bottle in my hand is suddenly light, and I realise that I'm up for my second drink already.

The speakers bing-bong back into silence and, released from the spell, everyone starts walking again. Richard and I soon reach the stern, and lean against a railing that overlooks a wall of lifeboats. To our left, the drop is sheer and the railings are just below chest height. It's crowded here now but if it were deserted it would be the easiest thing in the world for someone else, someone fragile, someone unpredictable, someone storing up fifteen years' worth of anger, to rush from behind and tip the other person into the churning water. It really wouldn't take much at all.

We spend a few moments watching the North Sea churn to a creamy yellow foam in our wake. A family of seals swims dangerously, stupidly, close to the ship, arousing in me sentimental thoughts of my own incipient family. All roads lead back to Laura.

'First picture of the trip,' I say, setting my phone to selfie mode. I hold it at arm's length and manage to capture the two of us, grinning over brown bottles of beer, the sea a blue blur in the background. When I try to send it to Laura, there's no signal. That's the problem with the middle of the North Sea; no phone masts.

'You can get Wi-Fi, but only in the lobby, although apparently you can pick up a signal on most of the top deck,' says Richard, whose hourly weather updates depend on knowing these things. On my land-lubber's legs, I climb a white-painted staircase to the next level; the metal banister is cold under my hand. Up here, it's deserted, and I can see why. The wind is sharp, loud and disorienting. I take shelter behind a steaming air vent, where it's warmer as well as quieter. It takes under a minute for my phone to lock on to the Wi-Fi, and I send the picture to Laura in London.

Looks like the world's geekiest ever stag do, she bounces back. I smile and pocket my phone, and turn to rejoin Richard.

Leaning on the railing with her back to me is Beth.

It's the hair I recognise first, that wild unbrushed mess that writhes in the wind like something alive. The image of her is so ingrained in my retina that I can process the rest of her in a split second. Right height, right curves, and those crazy patterned trousers that billow around her legs are just the sort of thing she would wear. The scratchy combination of sympathy and fear is back, stronger than ever. Here come the usual flashbacks, a time-lapse film from the moment we each met Beth to our last sight of her: the day-dark field, the courtroom, the stranger in our flat, the figure in the dust cloud, the moving image on a computer screen. The end credits; Laura, this morning, at the top of our steps, brushing imaginary cobwebs away from her arms before letting her hands drop to her belly. And now the final frame. I don't know how it ends. I don't have a plan. I've been waiting so long that I'm not ready any more.

II

LAURA
8 May 2000

Lachlan McCall died on the first May Day of the new millennium and was buried three days later. At the wake, Mac got so drunk he threw up all over Kit. We had to pick Kit's only suit up from the dry-cleaner's on the way down to Cornwall.

Jamie Balcombe's rape trial began on a Monday. That morning, Kit and I stood with our backs to the courthouse, looking down and around over the little city of Truro, the last in England. Behind us, a high Victorian viaduct curved a huge protective arm around the valley. Below nestled the modest cathedral, surrounded on all sides by the kind of red rooftops that Santa sails over in his sleigh. The crown court sat on top of a vertiginous hill; a terrace of little pastel cottages seemed about to slide down the slope and land in the river Kenywn, which gurgled over a thin weir. Feeling that the slightest tilt forward would tumble me straight down that hill, I gripped Kit's arm.

'They won't need us this morning,' he said, returning my squeeze. 'They've got to set it all up, and get Beth's version of events. That'll take till lunchtime, at least.'

Technically – of course, as ever – Kit was right. There was no chance of either of us being called to give evidence yet. The witness care service had told us that today would be about the prosecution setting out their case. Even Beth might not be called today and, as secondary witnesses, we weren't allowed in the courtroom until she had finished giving her evidence. I acknowledged the legal wisdom of this even as I flared against it. Nothing she could say would change my original statement. I knew what I had seen.

It was our first time back in Cornwall since the Lizard, and only our second time out of London. We had been conscientious students but the new challenge of building a career was a shock to both our systems. I was temping and Kit was a post-grad, having waltzed with his double first from Oxford straight into a doctorate at UCL, with a part-time sideline marking undergraduate essays; this wretched trip was the closest we were going to get to a holiday. Our hotel room was paid up until the Thursday, and our return train tickets booked for Friday morning. Meagre but welcome expenses were being paid. We had arrived late the previous evening and, after a sleepless night in a chintzy hotel with smugglers-cove paintings on the walls, we were still creaky from the long drive down from our little flat in Clapham. Just the two of us lived there. The original plan to get a houseshare with Mac and Ling had been scuppered by Juno, the baby they had unintentionally conceived a few weeks before the Cornwall eclipse, now two months old and with lungs the size of London.

Since the Lizard, or rather since we knew that the case would come to trial, we'd been mugging up on court procedure. My new knowledge swung me between hope and despair. I took some comfort knowing that so many rapes never even got tried; how ironclad a case must be for the Crown Prosecution Service to take it to court. Then I'd think about the conviction rates again, and feel a plunging despondency. The only constant was the image of a girl, covered in snot, face down in the mud, being assaulted, being *raped*. Somewhere between then and now I had stopped being scared of the word and armed myself with it instead.

'They'll probably let us go as soon as we've reported for duty,' he said as we approached the front door. I looked up at the faux-Romanic columns, pebbledashed then painted an ugly weatherproof grey. The day's listings pinned to an outside wall told us that R. v. Balcombe would be held in Court One, with Mr Nathaniel Polglase prosecuting, Miss Fiona Price defending and Justice Edmund Frenchay presiding. Reality pulled its focus sharp.

'I feel like we should stay anyway,' I said. 'Even if they don't call us.'

'I don't know if I can handle being cooped up in court after the last few days in funeral parlours and function rooms,' said Kit.

'Oh, love.' I had to keep reminding myself that he held more in his head than this court case. Knowing that Lachlan was on the way out hadn't made his passing any easier. I had hoped that the trial might give him, if not a break, then a change of focus. I was learning that Kit's reaction to anything challenging was to go into lockdown, letting his true attention out of the trap only for brief moments before retreating into silence.

'You ok to go in?' I asked.

'Yeah,' he said. 'Come on, let's—'

We were all but barged to one side by a knot of people in suits, walking in formation like bodyguards flanking the president. In the middle of this squad was Jamie Balcombe. A suit, a haircut, a shave; brown hair shiny, eyelashes glossy around wet blue pupils, he looked like a little boy in his dad's clothes. The swagger we'd seen before was gone; now he walked with measured precision, as though it was all he could do to put one foot in front of the other. He looked unnaturally directly ahead, making me think he must have recognised us, even though I'd deliberately dressed down in black, with my hair wound in a low bun so as not to draw attention. With my hair hidden I was insignificant, a lanky, slouching teenager.

At the door, the cluster of people thinned to a queue and then, as they passed through the arched metal detector, they became individuals. A racehorse of a woman with slicked-back grey hair bore no resemblance to anyone, but that man with a double chin and a signet ring could only be Jamie's father. Colouring gave away his brother and sister. The woman with the certain-age blow-dry, rubies in her ears and an elegant hand on Jamie's shoulder must have been his mother, and the wan redhead clutching his left hand must be another sister, or girlfriend. When the redhead took off her coat to pass it through the metal detector, I saw a diamond flash on her third finger. Not a girlfriend but a fiancée. She took the ring off and offered it to the security guard.

'No need, love, keep it on,' he said, but her fingers were clumsy and the ring slipped from her grip, landing on the floor with a little *chink* and rolling through the arch.

'Oh, Antonia,' said Jamie. The brother rushed off to fetch the ring; Antonia's fingers were shaking so badly that Jamie's mother had to help her replace it.

Kit and I hung back until the Balcombes had been processed. Then, as it was the only door we could see, we went after them.

'Hi,' I said to the security guard. 'I was just looking for the witnesses' entrance?'

'You've found it,' he said. 'Everyone has to come through here.'

It took a few seconds to sink in. We had just brushed shoulders with the man we were going to testify against. What about grudges, intimidations, revenge attack and threats?

'What if it's the mafia?' I blurted. Kit did a double take and the security guard laughed kindly.

'It's not exactly *The Godfather* round here, love. There in't nothing gonna happen with all these barristers crawling all over you all. You got a camera or a recording device in there?' I opened my bag to show him nothing but some Tampax and a notebook.

The interior of the court did little to put me at my ease. The architect must have been a sadist; surely the brief was to instil an air of solemnity without being too intimidating. This was an Escher-drawing of a place, with a layout of columns, colonnades and walkways that seemed to double back on themselves without warning.

Something awful occurred to me. 'What about the victims?' I asked Kit. 'Surely Beth won't have to come through that door?'

Kit looked aghast. 'I think they must, if that's the only security clearance. It doesn't seem right, does it?'

'It's a fucking travesty.'

We walked across a bustling atrium. The Greco-Roman theme continued here, columns supporting a ceiling the height of the building. One wall was obscured by tangled foliage in a massive planter that must have climbed 20 feet high, giving it a conservatorial air, or, as Kit said, 'It'll be like being on trial in bloody Center Parcs.' I appreciated the attempt to cheer me up, but could muster only the weakest of smiles.

It was half past nine and Court One had yet to open. Kit got us both coffees and for a while we watched everyone file in. Some looked up at the ceiling in wonder while others greeted the security guards by name. Most of the ceiling-gazers were instantly spirited away, enough in number that they must be the jurors. Carol Kent, the detective who'd taken my original statement, nodded at us from across the room, a little smile of encouragement momentarily softening the severe make-up.

'Let's go and see what we have to do,' said Kit.

Kent's greeting was curt. She's edgy too, I thought.

'I'll tell the prosecution you're here,' she said. 'You need to stay out of the public gallery until you've testified. There's a witness room but the complainant's already in there, so it's not ideal.'

Privately I thought that it would do Beth good to see a friendly face, and was about to say so when Kit said to Kent, 'What if I gave you my mobile number and we promised to stay close?'

The vertical line between Kent's eyebrows deepened. 'That should be fine, but let me check with counsel.'

She followed a sign pointing to the witness room.

'I feel like I should go and say good luck or at least hello,' I said.

'Did you not just hear what she said?' hissed Kit. 'There's all sorts of rules against that. You'd look like you were conferring with her. It's only going to jeopardise her case. What if it got thrown out, because of that?'

He was right, *again*.

Next to us, a glossy blonde was talking on a flip phone.

'Yeah, it's a rape this morning,' she said, with all the detachment of someone describing a routine dental appointment. A journalist; of course. '*Potentially* quite juicy. Our defendant's public school, lovely-looking boy. His dad's a big cheese, CEO of a FTSE 100, was in the same year as Prince Charles at Gordonstoun although they haven't kept in touch, more's the pity because a link like that would be *gold*. Look, I'll call you back at end of play, give you a good idea of whether it's got legs. It better had, to be honest. I've turned down a double murder in Liverpool for this.'

She dropped her voice as Carol Kent came our way.

'Mr Polglase says that's fine,' Kent told us. 'Keep your phone switched on, and don't wander out of signal.'

'Thanks.' Kit stood to leave.

The voice over the public address system was as Cornish as tin mines and pasties, long Rs raking in longer Os.

'Crown versus Balcombe, Court Room One.'

'I'm offski,' said the reporter. 'Talk later.' She slammed the aerial back into the body of the telephone and followed the Balcombes through the double doors. She was followed by another journalist with a geometric bob, a Press Association card on a lanyard around her neck.

Seconds later, the place was deserted, only us, a few court staff and a little blue butterfly flitting around the foliage. The usher from earlier looked at us and frowned. I felt suddenly under scrutiny. The atrium seemed to shrink around us.

'Let's get out of here,' I said.

'You've changed your tune.'

'Do you want to go or not?'

'All right, keep your hair on.'

We'd never spoken to each other like that before.

The pretty bit of Truro town centre doesn't extend far; within an hour we were already retracing our steps around the lanes. We wandered into the cathedral, pottered in bookshops and an art gallery. There was a museum but we decided to save that for if it rained. Lunch, in a pub at the bottom of the hill, was a baked potato, prawns swimming in mayonnaise and a half-pint each of a local ale called Bilgewater that tasted nicer than it sounded.

'I wish I knew what was being said in there,' I told Kit, grinding pepper over my plate.

'Well, you can't. That's the whole point of your testimony. That it needs to stand independently of the complainant's version. Otherwise it's worthless.'

I pushed my food away. 'I'm too tense to eat. What if our testimony isn't enough to put him away?'

'You're acting like the whole thing's on you. I've told you, there's a whole team of people who've been building a case for months. I wonder what forensics they'll have.'

We'd been through this before, too. It was an effort to keep my voice even. 'There might not *be* any forensics. It's just a case of who you believe.'

Kit shook his head. 'Words are just so . . . flimsy, aren't they? Like, you're always telling me I'm too binary. That I only think in black and white.' It was true: I nodded. 'But in this case, we're asking *everyone* to think that way and there's no evidence for any of it apart from words. How can you possibly secure a safe conviction that way?'

'What else is there?' I asked.

He couldn't answer, just looked pensive and took another inch off his drink. 'We did a pretend trial in sixth-form,' he said after a while. 'I had to give evidence. A hypothetical drug-trafficking case. I was bricking it even then just getting up, even when it was all fake and in the common room.'

'I didn't know that!' We were still at the point where new stories were both delightful – there is still so much to discover about you! – and an affront – why didn't I know this about you before?

'I take it Mac was the defendant.'

Kit's face flickered in a memory, then he laughed. 'Actually, he was the *judge*.'

'Fucking hell!'

'Our teacher had a sense of humour,' he conceded. 'Anyway, the point was, that once I actually did it, all the nerves went away, because I just said my thing, and that was all there was to it.'

I knew he was right, but it couldn't still the nerves that took away my appetite for food but gave me a thirst. If we hadn't been waiting for Kit's phone to ring, I could have sunk three pints. Fat raindrops hit the windows like stones.

'Museum it is, then,' I said.

But the museum was closed for the afternoon. We went back to the pub and played a game of pool on an uneven table. (I won, obviously: despite months of playing against me, Kit had yet to internalise the mathematics of trajectory and ricochet. Where I saw sport, he still saw applied projectile physics. He had a terrible poker face; you could always see the next move coming a whole minute before he picked up the cue.)

We wound up our game when we could no longer ignore the local teens looking longingly at the baize.

'It's three o'clock,' I said. 'She's probably in the witness box right now.'

'If they needed us back, they'd have called us by now. Let's go back to the hotel.' We handed over our cues.

'What are we going to do in a hotel room all afternoon?' I whinged, as I pulled on my coat. Kit smiled.

'God, things aren't *that* bad, are they?'

They weren't that bad, but even half a pint of beer hits hard on an empty stomach, so I just lay there making the usual noises while Kit went heedlessly through the usual motions. He finished in double time and was asleep in seconds, the release of orgasm a traceless Mickey Finn. I knew it would be an hour at least until he woke up. Outside, the rain had stopped and the rooftops of Truro glistened in the sun.

The clock said quarter past four. An idea came to me and I was back in my clothes before I could change my mind, out of the door and following the magnet pull to the courthouse on top of the hill.

12

LAURA
8 May 2000

I arrived panting at the court five minutes later. Knowing what to expect this time, and without the delay of the morning rush hour, I was through security in seconds. I bought a coffee to give me a sense of purpose but I hadn't even had a chance to blow on it before the doors burst open without warning and the court spilled back into the atrium.

The Balcombes were out first, swooping on Jamie as he emerged through a side door that I surmised must lead straight to the dock. The horsey woman I'd seen earlier confirmed herself as the defence counsel, wearing her gown and carrying her wig, talking not to Jamie but to his father. Their voices, bred to carry along corridors of power, echoed long after they were out of sight. Through the double doors emerged a harried-looking man in a brown suit. The tatty file in his hands was tied with red ribbon. He was walking almost sideways in his haste to get away from someone behind him. Hot on his heels was Carol Kent, arms folded, and behind her, at last, was Beth. Instinctively I pressed myself behind a white column, the better to look at her. Her appearance had been tamed – her curly hair and curvy body were in a tight plait and a tailored jacket respectively – but she was as vocal now as she had been mute at our first encounter.

'Are you fucking joking me?' she said. I couldn't quite place her accent. She turned to Carol. 'Can he do this?'

'I didn't say you weren't *allowed*,' said the lawyer. 'I said we don't *advise* it.' He was reaching for sympathy but grasping condescension. A red-faced man and a red-eyed woman joined the knot. His dark hair and her figure, heavy about the hips and the breasts, marked them as Beth's parents. She was dumpy and he was dowdy; they looked like the Balcombes' country cousins, or the serfs who worked their land.

'What's up, love?' said her father.

'They want me to go home now they've finished with me,' said Beth. She and her parents turned as one to Kent.

'I'm sorry, Beth, he's right,' said the police officer gently. 'Legally, you have the right to sit in the public gallery. But the thing is, juries don't like it. Especially in cases like this, where it's your word against his, it can look . . . vindictive.'

'I fucking *am* vindictive! I want to cut his dick off!' Beth gave a harsh, cynical laugh. Her father looked mortified; Carol Kent and the lawyer shushed her in unison, while her mother tried to put a comforting arm around her shoulder. Beth shrugged her off and pushed imaginary air downwards, visibly gathering herself, then continued at a normal register. 'I've just had to relive what he did to me while he sat there giving the jury Bambi eyes. Don't I get to watch *your* lot do the same thing to *him*?'

'I do understand,' said the lawyer. 'I'm sorry; it does seem weighted against you. I hate having to advise it. I can't stop you if you really want to attend. But I wouldn't be doing my job if I didn't tell you that we stand a better chance of winning the case if you're absent. It's just how juries behave.'

'Why didn't you tell me this before?'

Carol turned away from me and said something I couldn't quite catch.

'You thought it would put me off my *stride*?' Beth looked suddenly punctured. 'Well, maybe you're right, maybe you're—'

She broke off mid-sentence and walked away, her back artificially straight; in the single spasm of her shoulders I recognised a proud woman trying not to cry. Her parents bore her off down one of those mysterious hallways, one at either shoulder.

'Fuck,' said Kent to the lawyer. 'That backfired.' She puffed out her bottom lip. 'I'll sort this.'

Across the room, almost hidden inside the little grove of leaves, I saw the blonde journalist from the morning, punching numbers into her phone. Even though no one was watching, I circled the whole atrium before pressing myself against a pillar. She was just visible between two huge ferns.

'Hi, it's Ali. Good news. I can definitely place this. It's going to be a classic he said/she said, textbook case decision by jury. Half the female jurors are already in love with him. Mm, mm.' She nodded into her phone. 'Bit weepy but holding it together. She made eye contact with him, which they don't always, so she's a feisty one. Reckons he chatted her up the night before, she gave him the knock-back and he wouldn't take no for an answer, yada yada yada.' I wanted to knock the stupid phone out of her hand. 'The usual. It always sounds convincing until the defence cross-examine the victim and they *shredded* her.' I was left to imagine the distress this Ali had witnessed, and with physical effort I kept a lid on my fury at the way she was boiling it down to the stock of a story. She looked through her notes. 'There's a history of depression, which is never good. Yeah. About the only good thing you can say for the poor cow is that she hasn't got form for casual sex. Defence couldn't even pull an old one-night stand out of the hat.' There was a question on the other end. 'Well, we'll find out about *that* when he goes into the box, won't we? But, you know, I've seen dark-alleyway rapists get off scot-free, so . . . Anyway, tell the boss that this is me for the week. I'm grooming the fiancée for the story. If he walks, she's a heroine, standing by her man. If he goes down, she's a wronged woman. It's good copy either way. She's very pretty, too, front-page pretty. Make sure the picture desk call in some good shots of her.'

My indignation bubbled like lava; to stop the eruption, I went to the toilet to run my wrists under the cold tap, a trick my dad had taught me to keep my temper. In here the ceilings were abruptly lower. The court bathroom was shabby and had seen better days. One cubicle was occupied but the other was swinging wide and vacant; people don't hang around a courthouse, I was fast discovering. It empties as quickly as it fills. I put my hand under cool water and waited for my anger to die down.

The cubicle door behind me swung open with a creak, and I found myself locking eyes with Beth.

I heard Kit, and Carol Kent, in my head; you shouldn't be here.

The right thing to do was to walk off, but this was a human being, and one who had just been 'shredded' in the witness box. My smile was all I had to give her.

'Hey,' I said softly.

'It's you,' she said nervously, but she did smile back; then she grew serious, double-checking the empty cubicles, and shot a glance towards the door. 'This is against the rules.' She said it more in the spirit of *this will jeopardise my case* than moral conflict.

The door to the corridor clicked open and we both jumped but it was only the breeze.

'It's ok, I know you can't talk about it,' I said. I felt a little flare of pride in my correct behaviour, and anticipation of Kit's approval. 'I heard you talking to the barrister about not coming back tomorrow.'

'Who? Oh. He's not the barrister, he's the CPS caseworker, barrister's dogsbody.' Beth took off her jacket. Underneath she was wearing a black shift dress with no sleeves. I started; her arms, which I remembered as soft and white, rippled with muscle and her shoulders were as broad as Kit's. You didn't need a degree in women's studies to realise that this was a reaction to the rape. The bulging biceps were as damning as scars. I watched her wash her hands, soaping them past the wrist. Muscles twisted like ropes under her skin, and the awful thought rushed at me; I hope she kept her jacket on when she was giving evidence. She looks too powerful now. She doesn't look like a victim.

I hadn't even been into the courtroom yet, and already I was thinking like them.

'It's probably for the best.' She rinsed her hands and shook them dry. 'Because this way I don't have to run the gamut of him and his bloody family every day. They were next to us in the car park. Have you seen what they drive? Massive fucking Jaguar.' She was talking to her own reflection rather than me. 'And Carol's going to keep me posted.' Beth leaned, almost fell, forward, pitching her head until her brow was pressed against the mirror, then her palms. 'I don't think I can live through the next few days, not knowing what's being said about me in that dock,' she said. 'He's going to get up and lie through his teeth and they're going to believe him. *Look* at him. He looks like butter wouldn't melt. I don't think I can survive it going his way.'

The urge, somewhere between sisterly and maternal, was a warmth spreading from my heart throughout my body. It was such an instinctive thing that I didn't know I was making a decision, let alone what I was setting in motion.

'Oh, Beth.' I put my hand on the small of her back. 'It will go your way, because he's guilty. I know what I saw.' She smiled weakly at my reflection. 'But if it doesn't,' her spine went rigid under my palm; 'I mean, it will, but if it doesn't, and you need to talk to someone who believes you, give me a call.' I rummaged in my bag for something to write on and found only a work card, the generic business card they gave all the temps. I wrote my mobile number on the back of it.

Beth put her hands on my upper arms and gave me a watery smile. I could see the effort of holding back tears in the tiny convulsions of her lips. 'Thank you,' she mouthed. She took a deep breath in and a long breath out. 'Ok, I've got to go. I need to talk to Carol and my parents are in the car park. It's a long old drive back to Nottingham.'

When she had gone, I briefly laid my own forehead against the imprint hers had left on the mirror. I had done it. I had wanted more than anything to know what they had said in court, but I hadn't put her in the impossible position of asking her. Kit would be proud when I told him.

I trotted down the hill, pulling out my bun as I crossed the little bridge over the Kenwyn. I felt, if not light – the morning would bring me to the witness box, after all – then light*er*. Whatever else happened, Beth knew now that I was on her side. I hoped that, whatever was to come, she would always be able to find solace in that.

Kit was just waking up when I twirled the key in the lock; he had the surprised look of someone who's woken up in the middle of the day, puffy around the eyes and lips. He pulled back the covers and I slid in beside him, my clothes cold against his hot skin.

'Where'd you go?'

I don't know what made me say it.

'Out for a walk. Along the river, just around.'

It was my first lie.

13

KIT
18 March 2015

My instinct is to run back to my cabin. Do what Laura does and make a list of what I need to say. Fifteen years, it seems, is not long enough to prepare for this conversation; I need another hour.

I should run. And yet I am paralysed, my boots apparently glued to the deck, my eyes fixed on Beth. Has she seen me? She's done this before, kept her back to me when she *knew* I was watching her. I always reasoned that when someone has taken so much of your power away, you'll work with what little you have left.

I'm not close enough to touch her but I can see the droplets that sit on top of her hair like tiny pearls. Move, I tell myself. I breathe in just as a huge wave hits the liner's prow and inhale deeply the spray of brine as it hits me in the face. It's up my nose and in my throat, making me splutter out loud. Beth whips around to see what the noise is. She's been soaked, too, her hair flattened to her head on one side. It takes a second to understand that it's not her, not at all. This woman looks my age from behind but she's got to be nearly sixty. The hair is dyed, it's obvious now; it's as dull as boot blacking. Her jawline is loosening and there's a Mr Punch hook to her nose and chin, nothing like Beth's silent-movie cupid's bow and arched eyebrows.

'That's what you get for standing too close to the edge!' she says, laughing and wringing her sodden curls. I return a watered-down version of her smile, and raise my empty bottle to her. I'm psyching myself up to make small talk when an announcement comes over the public address system. It's a different voice; female, nasal, more suited to a department store than a ship full of astronomers.

'Please make your way to the ballroom for welcome drinks prior to this evening's entertainment.'

'Can't miss this!' says my new friend. 'Although I'd better go and towel off first.' I wave her off, knowing from all those years working on Laura's hair that she'll have her work cut out with those tangles later.

It's only after she's gone, and I'm alone on the deck again, that I look down and see my Swiss Army knife in my fist, the longest blade extended, almost as though someone has put it there without my knowing. It starts to tremble in my hand. I have absolutely no recollection of taking it out of my pocket, let alone pulling out the blade.

I learned a long time ago that the moment you start to think about the logistics of an act – even if you're only daydreaming ways you might go about it – you've already crossed a line. This, though, is different. This time action has preceded thought.

My heart runs at full pelt.

'Chris!' Richard calls from the deck below.

I fold my knife carefully back into my pocket. The dark-haired woman can't have seen it, or she wouldn't have been so friendly, she would have run screaming down the deck. 'I'll catch you up!' I shout back. I don't want to miss the introductory lecture but I need to let my pulse return to normal before I go back in among the crowds. I roll the bottle across my forehead from side to side, pressing it against my temples.

I have carried my Swiss Army knife on every trip since I was twelve, but it is only now I recognise it as a weapon rather than a tool. It's the loss of control – of awareness – that frightens me most. Perhaps I have always had a plan, only I never dared admit it, even to myself. In one involuntary action I have unlocked the fear that's been dogging me since I left London. Maybe it's not Beth I need to fear but what she might unleash in me.

On the upper deck, I squint at my phone until the Wi-Fi signal appears. I hit the tiles on my phone one by one, checking for updates on the blogs and chatrooms. Even as I scroll through the Facebook page I wonder what I'm doing. I don't really think Beth would be mad enough to post anything online – after the drubbing she had on the internet, she's hardly likely to court publicity, and in any game of cat and mouse both must be silent – but it's possible that someone's caught her in the background of a photograph. I don't know what I'm doing: only that I am compelled to do *something*.

ShadyLady on Facebook has written a three-word update: *Touchdown in Tórshavn*, accompanied by a photograph of the red house in the harbour that's no use to me. It probably isn't Beth, but I wish she would post just one selfie so I could stop the constant checking.

The remaining posts are all about the weather, and nothing has changed on that count.

I swipe as fast as the connection will allow. At last, on a blog so obscure I nearly didn't bother to add it to my list, a picture catches my eye.

On the Faroes, toasting the gods of sunshine with friends old and new, says the caption, beneath a photograph of six men and women sinking tankards of beer in a darkened bar. Their faces are hidden by the bases of their glasses, six frosted circles obscuring identity. The youngish woman on the edge of the shot has white skin and black curly hair tied in a topknot.

It is impossible not to sketch Beth's face in the blank.

In the comments someone anonymous has written

Same time, same place tomorrow night?

There's a thumbs-up emoji in reply, and below that, the blogger says

And every night till totality! followed by a row of little tankards and cartoon suns.

I stare deep into the background of the photograph. There's distinctive stone cladding around an open fire and on the wall, a huge watercolour of some kind of elk. It's not the kind of interior you'd find more than once.

I could find that bar, I think. The knowledge that, if I wanted to, I could sit there and wait for her, gives me a sharp tingle, like I've tested a battery with the tip of my tongue. As the shock recedes, the practicalities present themselves. Do I really want to spend the next two days haring around from bar to bar? I'm acting as though the mere opportunity, combined with Laura's absence, obliges me to confront Beth. Exhausted, I shut down the picture and close my eyes.

14

LAURA
18 March 2015

Mac's bright green ectoplasm drink tastes nicer than it smells. I sip it through a straw in the kitchen, listening to the woman on BBC London hosting a phone-in about the great eclipse goggle famine of 2015. Yesterday, the newsagent on Green Lanes had a sign in the window saying NO ECLIPSE GLASSES HERE. SKY AT NIGHT MAGAZINE SOLD OUT. If the weather was better, I could make some money by selling some of ours. We've got dozens of them, from flimsy souvenir freebies to professional-grade polymer in strong plastic frames. According to the radio, over-protective parents are panicking London's children will all go blind, like in *Day of the Triffids*. One teacher is even keeping the poor kids indoors with the windows closed. Callers are frothing at the mouth about the health and safety culture. I don't know why the presenter is feeding the frenzy. She must know it's going to be clouded out. But I don't suppose that makes great radio. I hover over my iPad, which I keep online, much in the same way that Mac keeps a bottle of whisky in the house as a test of his sobriety. Sometimes he even opens it and smells it. The difference is, you can live without whisky, but you can't survive in London in 2015 without the internet. I check the weather quickly; it's still touch and go. Kit could yet get his eclipse. There are more detailed reports available but that means opening up the internet properly, so instead I put the iPad face down before I'm tempted to look at anything else. One click leads to another. Images slide in unbidden online and as soon as you start searching anything to do with eclipses there's an outside chance *the video* might find its way onto my screen.

The video is an adult version of a child's worst nightmare.

When I was about seven, I became frightened, properly phobic, of a picture in a book. There's a chapter in *Charlie and the Great Glass Elevator* by Roald Dahl in which one of the grandparents ages by hundreds of years in a matter of moments. The accompanying pen-and-ink drawing of the lined and sunken face scared me so much that the first time I saw it I wet myself. I couldn't read the book again until Dad went and bought me a different edition with new illustrations. I called it *the picture* – 'I've had a nightmare about *the picture*,' was all I needed to say when my screams called him into my bedroom in the middle of the night – and even now, when photographs have played such a significant part in my life, it remains *the* picture.

It's the same with *the video*. I know it frame-by-frame. (I've thought a lot about memory over the years; how some events imprint themselves clearly and never fade, while others are smudges even as they happen. Take the letters Jamie sent me, for example. Even though I didn't keep them, I expect I could recite every single one.) I could probably direct a reconstruction of it, from the fire-juggler who opens the film to the sweeping shots of the lush Turkish valley. I could describe perfectly the music, a classic trance beat overdubbed with wild wailing in that weird Eastern atonal key that was somehow unlocked to Western ears by the beat and the chords. And I could tell you that at ten minutes and fifty-one seconds into the video, a dark-haired girl with a photograph clutched in her hand – the one Ling took at our graduation ball? Or the one Beth took herself, the warning I chose to ignore? – interrupts twelve different people, drawing their eyes from the sky to the picture, surely asking if they have seen these people, do you know them, can you tell me where I might find them. Her whole part is only maybe twenty seconds long and it's all in the background, then the camera pans up to the sky as totality hits. But it's enough.

I know all this; it's a perfect sequence that plays itself out in my dreams. Why, then, am I so scared to watch it again?

I think I know the answer. While it's all in my head, I can pretend that's the only place it exists. I can file it away with the anxiety attacks and the rest of the paranoia; a product of my over-sensitive, over-working imagination. But when I watch *the video*, it is all too real. It happened.

Why didn't I see it? Looking at her now, her problems are etched on her face; you can tell by the way that strangers are looking at her. People who've never met her recoil from her intensity so why, when I spent so much time with her, didn't I see it until it was too late?

15

LAURA
9 May 2000

The witness room had a stale, sugary smell of stewed tea and biscuits that had been in their tin for more than one trial. I was due into the witness box first; Kit, who had seen only the aftermath, would follow me. Even though we had spent every night since last August together, we were forbidden from discussing the case. Carol Kent had told us not to speak at all in the break between my evidence and his, so we held hands in silence while the witness care officer, a sprightly blue-rinser called Zinnia, filled us in on the importance of wigs as predictors in English criminal justice.

'I can tell who's going to win just from looking at the wigs,' she said. 'You'd reckon that a successful barrister would have a nice tidy new one, wouldn't you?' She paused meaningfully, withholding the rest of the information until we both nodded. 'Wrong!' she said in triumph. 'The top barristers have really ratty ones, hundreds of years old, some of them are. It's a mark of quality. I wouldn't want someone in a brand-new wig fighting my corner.'

At half past ten, Zinnia walked me down a carpeted corridor and into Court One. Soft carpet swallowed my footsteps. The court looked more like a little cinema, with its royal blue carpets and tip-up seats, than the classical, wood-panelled rooms I was used to seeing on television. A digital clock on the clerk's desk told the time to the nearest second. The biggest shock of all was the public gallery; I had thought they would be up in balconies, but they were just *there*, behind a long table, in the far corner; no cordons, no barriers, nothing. Anyone mad enough could sprint from one side of the room to the other in two seconds. Jamie, behind glass in the dock, was close enough that I could see the stripes on his old school tie.

If the courtroom disappointed, the judge did not; he was the archetypal television judge, a Rumpolean wine-red face craggy under his powdered wig. The jurors looked me up and down. Apart from one Sikh man, everyone was white. In the front row I saw a stern professorial man, a mumsy woman with reading glasses on a beaded chain round her neck, and a very young man in an England football shirt whose tattoos crept over his collar.

Unprepossessing as they looked, they had a huge advantage over me. They had seen Beth's testimony. If only there was some telling detail she had shared that I could use to rivet my account of the rape to hers, to make watertight both our truths.

I walked to the witness box on legs of jelly and although I was offered a seat, I stayed standing. The air was thick and still, with a solemnity that must act like a truth serum on the guiltiest of people. Surely Balcombe couldn't continue to maintain denial somewhere like this. He blinked at the judge and jury, long eyelashes batting away the allegations.

The prosecuting barrister, Nathaniel Polglase, was thirty-five or so and his horsehair wig looked box-fresh, each curl a tight glossy roll. Behind him, the CPS caseworker was wearing yesterday's clothes. I knew that they weren't, technically, Beth's lawyers, but they were her only chance, and I felt a little spike of preemptive disappointment. To the judge's left, Jamie Balcombe's defence barrister, Fiona Price, was slick as steel. Her wig was tatty but she wore it well, and she looked so good in her gown I couldn't now imagine her wearing anything else. I dared to glance at the public gallery; Jamie's family sat two rows deep. On the press bench was the blonde journalist, Ali, another female journalist, a youth of about sixteen in a clip-on tie and a middle-aged man evidently on the verge of sleep.

I took the secular oath – in Kit's eyes, anyone who swore on a holy book ruled their evidence inadmissible by reason of extreme stupidity – and parroted my promise with my eyes on the huge crest on the wall behind the judge; in dull gold relief, the lion and the unicorn battled for the crown. Tug of war, tug of love? You could read it either way.

'Thank you for coming, Miss Langrishe,' said Nathaniel Polglase. His accent was local and he pronounced my name to rhyme with *language*. 'I wonder if you could, before you begin, let us know a little about yourself. Career, education, that kind of thing.'

'I went to a comprehensive in Croydon, in Surrey. Ten GCSEs, three A levels,' I said, feeling like the teenager I'd been last time I'd listed these qualifications. 'And I got a 2.1 in Sociology with Women's Studies from King's College, London, last year. At the moment I'm working as a temp in the City, in advertising sales.' This abridged CV was evidently all he needed from me as then, with a series of questions, he coaxed from me a version of events that was pretty much word-for-word the statement I had given to the police back in August. It sounded flat, weird, rehearsed; which of course it was. I kept looking up at the jurors to gauge their reactions. The man with the tattoos wasn't even watching me. Somehow my words did not seem as powerful in the sterility of the courtroom as they had in the police cabin. The jurors didn't seem to be engaging with me. I'm failing Beth, I thought, as I approached the part where I picked up the purse. I'm losing the jury and there's nothing I can do about it. When I had finished, it was all I could do not to ask, 'Is that all you want from me?'

I had expected some kind of break in between prosecution and defence, but Jamie's counsel was on her feet before Polglase landed in his seat. She had the jury's respect instantly; she had my respect instantly.

'Miss Lang-*reesh*,' she said, the correct pronunciation drawing attention to Polglase's mistake. 'When did the complainant tell you she had been raped?'

At first, I didn't see where she was going. 'She didn't. Not in so many words, but—'

'Not in *any* words. Who was the first person to *suggest* it was rape?'

My blood spiralled in my veins as my understanding caught up.

'Me, I suppose, but I wasn't suggesting, I was just saying what I'd seen.'

'So the complainant Miss Taylor did not say that it was rape, or that she had been forced, or anything, before you called the police?'

'Well, she wasn't talking then.'

'So she said it was rape when they came?'

I only understood she'd been trailing searchlights around me when they shone directly into my eyes. How had it gone so wrong, so quickly? 'No, but—'

'At no point did the complainant tell you that she had been raped. In fact, you arrived at this conclusion by yourself.'

I put my hands flat on the wood of the witness box. 'Actually if anything *he* was the one who said it first.' I glared at the dock. 'He said "It's not what it looks like" when he was pulling it out. I hadn't even spoken at this point. So if anyone was on the defensive, it was him.'

Fiona Price hooked her thumbs under the lapels of her gown and raised a skinny eyebrow at the jury. 'So he said that it wasn't rape, that it wasn't what it looked like?'

'Because he knew that it was.'

I grew hotter, as though someone was turning up a dial inside me; it was a warning sign, a needle trembling in a red gauge.

'You haven't answered my question. He protested his innocence, didn't he? "It's not what it looks like." Those were his words, according to you, weren't they?'

'Well – yes – but . . .' Of course he did, I was going to say. That's what people do when you've caught them out. That's what little children do, deny that they've been in the biscuit tin even when their hands are sticky and there are crumbs on their lips. By the time I'd formulated this thought, Price was already well into her next question.

'In fact, the defendant Mr Balcombe was telling the *truth*, protesting a *genuine* innocence, seeing your *obvious* misinterpretation of the situation and jumping in before this misunderstanding had a chance to escalate, wasn't he?'

'No.' My armpits were clammy, and I was glad of my dark dress.

'You'd already made up your mind, hadn't you?

'It's not a question of making your mind up. You know it when you see it.'

'What you interrupted was consensual sex, wasn't it? Vigorous, intense sex, yes, but something that took place between two *consenting* adults, that you *interrupted*?'

'I know what I saw.'

'You saw the last moments of sexual intercourse, Miss Langrishe. Were you there at the point of penetration; the point at which consent was *allegedly* withheld?'

My hairline was soaked. I put up a hand to wipe my brow, thought better of it, and wondered if the jurors could see the resultant trickle down the side of my face. Fiona Price certainly could; she traced its route with a smirk.

'You know I wasn't.'

'So you agree that the only two people who can *possibly* know whether consent was given are the participants; that you can't know?'

I might not have had the advantages of legal training or education, but I had right on my side. 'No, I don't accept that. You didn't see their faces. He was so angry. And she was all, she had mud around her mouth where she'd been crying and dribbling, and snot all over her face.'

'I must remember, in future cases, that an unwiped nose is evidence of rape.'

Tears began to prick, not just for my own embarrassment, but for Beth. If they were doing this to me, what must she have gone through?

'*Really*, Miss Price?' The judge's warning shot landed short of its target. The mumsy woman in the jury smirked.

'Your Honour.' Fiona Price dipped her head in apology for a count of three. When she raised it again, her eyes were lasers.

'Miss Langrishe, did the complainant display any sign that she had noticed you when you first interrupted them?'

I thought back to Beth's thousand-yard stare. How long before she'd registered me? It was impossible to say. Time had stretched as I took it all in, the way they say it does during a car crash.

'It took a few seconds.'

I could tell by Price's expression that I'd said the wrong thing.

'She was lost in the moment, wasn't she?' crowed the barrister. The words *lost in the moment* described something so other from what I had seen. She had locked herself out of it. She was trying not to be there. And how dare Price reframe it as the throes of pleasure? How was she allowed to do this?

'If anything she was frozen in fear. He was hurting her!' My voice was shrill.

'Miss Langrishe,' interrupted Price, but I kept talking without meaning to; it was almost as though I could see the words spilling out, a script scrolling in front of me, and I had to read it even though I had no idea what was coming next.

'He had great big handfuls of her hair in his fists, and she was saying, please, no, don't.'

Nathaniel Polglase's head shot up so quickly his wig bounced. The scroll of paper in my mind's eye abruptly trimmed and framed itself, my words hanging in the air for all to see. Fiona Price seemed to glide towards me.

'Forgive me, I didn't catch that last sentence. Could you repeat it, for my benefit?'

I had the sense of something huge falling away behind me. 'She was saying, please, no.' I said it as powerfully as I could, but my body told the truth. My face streamed with sweat. I don't know what it looked like from the outside, but I felt as though someone had squeezed a warm, wet sponge over my head.

'Allow me a moment, Your Honour,' said Fiona Price. 'I'd just like to skim-read the witness's original police statement.'

She put on reading glasses and appeared to give the document before her serious attention, as though she were reading it for the first time, even though I was sure she had memorised everything in front of her and this pause was stagecraft. 'Can you remember whether you mentioned when questioned by my learned friend for the prosecution that the complainant used the word no?'

'I didn't.' She's humiliating me, I thought, realising only as she reached out to pass me the witness statement that she was too clever for that; she was letting me humiliate myself.

'I wonder if you could read your witness statement and tell the jury where you mention that the complainant said the word no.'

My eyes skidded unseeing across the page, even though I knew the words wouldn't be there.

'That is your witness statement?

'Yes.' There was a kind of relief in finally agreeing with her.

'When was that statement taken?'

'Straight afterwards. In the police cabin.' After he raped her, I wanted to say, but I knew that would only show my bias, and I'd done more than enough damage already.

'In the immediate aftermath of the alleged rape,' said Price. 'Please read out loud the sentence at the start of the third paragraph down. Page 110, Your Honour, third paragraph down.'

I sounded like a backward child reading aloud in class. ' "She didn't say anything, it was more of a whimpering sound." '

' "She didn't say anything, it was more of a whimpering sound"? That's what you told the police immediately after it happened?'

'Yes.' The lump in my throat must have been visible from the public gallery.

'Now, there's a *world* of difference between a whimper and a word, you'd agree?' said Fiona Price.

'Yes.' It came out in a whisper. My eyes swam; tears, if I wasn't strong, would come with the next blink.

'You misinterpreted a moan of pleasure, didn't you?'

'No.'

'And if she'd said a word, you would have told the police, because *words* are not so easily mistaken, or forgotten, are they?'

I didn't shake my head as firmly as I wanted to, in case I dislodged a tear.

'And yet you have just told this court that you clearly heard her say, please no. You agree that only one of these statements can be true, Miss Langrishe?'

My chance to put things right was slight – it was like hoping that a single thread could bear my bodyweight – but it was all I had to grasp.

'I mean, I heard her whimper, please, no,' I said, in something of a whimper myself.

Fiona Price's eyebrows all but vanished under her wig. 'A clever girl like you *must* be aware how important those words would be in a case like this, yet you overlooked this at the time of giving your statement, mere *hours* after the alleged assault?'

I could only shrug. Staring at my feet, I understood why people talked about wanting the ground to swallow them whole. Come on, I thought, looking at the fuzzy blue carpet. Part. Show me the abyss. Put me out of my misery.

'And you didn't remember this mere minutes ago, in this very witness box? And yet it suddenly all comes *flooding* back to you under cross-examination?'

'She did, she did say please, no.' She might as well have, I told myself. She might as well have. I wish she had.

Price let my words hang in the air for a few seconds, then changed tack.

'How would you describe the atmosphere leading up to the alleged assault?'

I cooled a little, glad that the spotlight had swung away, even knowing that the respite would only be temporary. I swallowed the lump in my throat.

'Licentious?' asked Price. 'Hedonistic?'

'No. It was quite quiet for a festival,' I said. 'But it was happy. Peaceful.' She can't get me on that, I thought.

'Anything goes? You'd just seen an eclipse, the music was on, everything was heightened, wasn't it?'

'That doesn't excuse what he did.' I bit back a smile of triumph. At last I'd got one over on her. Or so I thought, until she went on and revealed that it was my state of mind she was burrowing towards, not Jamie's.

'In fact, the eclipse was, I believe the phrase is, clouded out? You didn't see it?'

'Yes.' I didn't realise, but I was laying a path towards my own further humiliation.

'Visibility was poor that day. Was there a sense of anti-climax, after all the anticipation of the eclipse?'

'Yes.' Confidently, unwittingly, I set another stone.

Fiona Price stroked her chin. 'Are you prone to getting caught up in drama, Laura?'

'What?'

'Making things up? Struggling to tell the difference between reality and your imagination?'

I dug my fingernails into my palm, furious at myself.

'No,' I said. 'I'm just telling you what I heard.'

'What you *heard* seems to vary from one minute to the next.' She smiled, almost in sorrow; as though she deeply regretted my unreliability. 'What you *saw* was two adults making love, wasn't it?'

I shook my head hard. 'If you'd seen what I saw, you'd be ashamed to use that phrase in this context. This was not "making love". I saw a *rape*. I saw it in their faces.'

Price regarded me with what looked like pity. 'Do you know what confabulation means, Miss Langrishe?'

Erin Kelly

The droplet of perspiration that had been inching its way down my face splashed on to the ledge in front of me. 'Of course I do.'

'Forgive me. You're a highly educated young woman; an upper second-class honours degree in Sociology and *Women's Studies*, lest we forget.' She tilted her body towards the jury. 'Jurors, for those of you not familiar with the term, and to remind those of you who are, confabulation is merely a long word for jumping to conclusions; when someone genuinely comes to believe in something that fits a pet theory of theirs. Are you a feminist?'

I hit back with the hardest argument to rebut. 'I believe women and men are equal.'

'In your studies, you've read the theories that all men are rapists?'

'With respect, that's a bit of a cliché.' The jury all but crackled with hostility; don't be a smart-arse, Laura, I told myself. Price herself was unruffled.

'But you've read them?'

'Of course.' I tried to crowbar some politeness into my voice.

'And you assumed this was the case here, didn't you?'

'No. I saw that man rape the complainant.' The mental picture of what I'd seen burned vividly in my mind's eye. The urge to cry was back, stronger than ever, as Price turned on me.

'It is your *confabulation* that you saw a rape, isn't it? The work of an overactive imagination so uncontrolled that you can't even tell the same story twice?'

'I saw their faces,' I said, but it was a choice between talking and tears, and it came out as a whisper. In any case, my words would have been drowned out by the clear and strident tones of Fiona Price saying that she had no further questions for me.

'No re-examination from me, Your Honour,' said Polglase through clenched teeth. I couldn't look in his direction. I'd gone from his star witness to his saboteur in the space of . . . I looked at the clock and was staggered to find I had only been up there for twenty-five minutes.

As I was led from the witness box, I heard the judge saying that
we would break for lunch, his words warped as though bubbled
through water. I was probably supposed to wait for Carol Kent
outside but I fled the courtroom before the atrium could fill again.
I was half expecting uniformed officers to arrest me for perjury at
the door. Outside, I took deep lungfuls of air as the Truro skyline
swam in and out of focus. I felt like hurling myself down the hill
and into the river. Fiona Price had read my mind. She had seen
what I was doing before I had. I had blown the whole thing. The
only thing that had – not gone well, as such, but not been a disaster – was that Kit hadn't seen me perjure myself. I wasn't who he
thought I was. I wasn't who *I* thought I was.

16

LAURA
9 May 2000

I was allowed into the public gallery now that I had testified. During that lonely lunch hour, though, as I paced around the drizzly courtyard, I thought seriously about not returning to court at all. But how would I explain my absence to Kit? He would know that something was wrong. So, at two o'clock, I filed behind the Balcombes into Court One. From the witness box I would have said that Jamie's entourage filled the gallery but still they managed to arrange themselves so that the seats either side and in front of me were empty. His mother – wearing emeralds today – actually cringed whenever I moved, but this only made me sit taller and gave me a new resolve. Fronting it out like this might actually strengthen my case in the jurors' eyes. It was the perfect bluff; who would lie in court and then put herself back under scrutiny?

Kit naturally took the secular oath. He hid his nerves with a bluster that must have looked to strangers confident, pompous, even. Like me, Kit was introduced by way of his education – a couple of jurors, including the mumsy woman and the Sikh man, actually nodded their heads in approval at the mention of his double first from Oxford – and then, like me, he recounted the events of that August morning. Nathaniel Polglase had no further questions, but Fiona Price was eager. 'Can you take me through, once again, the seconds before Mr Balcombe departed the scene?'

Kit nodded. 'Laura was shielding the girl, and she, that's Laura, was in a stand-off with the defendant. He was trying to laugh it off with me, but it wasn't very convincing; it was sort of forced laughter. He only took it seriously when Laura said we had to get the police along.'

'Did you give him a chance to justify his departure from the scene?' Her tone was different with Kit, free from the rhythmic, hectoring inflections that had characterised her cross-examination of me.

'No, but he didn't try to.'

'I didn't really get a feeling for the defendant's pace in your original statement. Was he walking like a man who was trying to escape harassment?'

Kit gave it some thought. 'He didn't *run*, but he definitely picked up his pace when he saw me coming after him, and then he was swallowed up by the crowd. I followed him for a bit and then another crowd merged with us and it was hopeless.' He looked stricken all over again at the memory.

She folded her arms underneath her gown. 'What were you going to do if you caught up with him?'

'Honestly? I don't know. It was all so in the moment. I suppose I thought I'd make some kind of citizen's arrest, although I wouldn't have had a clue how to go about that.'

'So. You didn't see the intercourse?'

'No.' Kit's voice was dispassionate, disconnected and utterly authoritative.

'You barely saw the victim?'

'That's right.' He could have been answering a survey, he was so collected. I'd previously criticised Kit's ability to divide fact from emotion. Now I only envied him. Why can't I do that, I thought at the time. I realised later there was one crucial difference; he was telling the truth.

'And you didn't bother to ask the defendant why he was running away?'

'No.'

'So essentially your testimony is just an extension of your girl-friend's, isn't it? You were acting on the basis of what she *decided* he'd done?'

'I trust Laura's judgement.' said Kit evenly. Guilt wormed inside me.

'Can you state, under oath,' pressed Fiona Price, 'that the sex was non-consensual?'

'Of course I can't,' said Kit. 'You can't swear to something you didn't witness.' His smile disarmed Price into a worried frown, the first genuine expression I'd seen on her face all day. Kit had, with the counterweight of his own bold truth, undermined her with her own question. A ripple of annoyance ran through the assembled ranks of Balcombes at my side. Precious stones and expensive watches rattled as arms were folded and heads shaken. They're going to give her hell for that, I thought.

I changed out of my dress as soon as we were back in the hotel room. Kit sat cross-legged on the bed, still in his suit, a tourist map of Cornwall spread out before him, gazing unseeing at the snaking coastline.

'I don't think I realised what a fucking minefield it was till I was up there in the witness box,' he said. I didn't know if he was refer-ring to sexual boundaries, the criminal justice system or entering into a debate about either of those things with me. 'I mean, when they were asking me about the way he ran off; you know, is he guilty or innocent? I'm glad we don't have to go back again.'

I froze with my dress in my hands. We still had three days left in Cornwall: I had assumed we would continue to follow the trial.

'We could go and see Goonhilly Downs tomorrow,' said Kit. 'You know, that place with the standing stones and the satellite receivers that I told you about back in the summer?' I kept my back to him as I hung up the dress and smoothed it down on its hanger in the wardrobe. 'Ok, not quaint enough for you. What about lunch in St Ives, then?' I still didn't say anything. He flopped backwards and the bedsprings sighed with him. 'You want to stay here and see out the trial.'

I turned to face him. 'Just long enough to see him in the witness box. I didn't get to see Beth but I want to see Jamie go through it. I want to see him deny it and I want to be there when he breaks.'

Kit didn't look convinced. I tried to appeal to his methodical nature. 'It's the police and doctor tomorrow. You might feel better if you sit through all the forensic stuff.'

'But you said yourself that it might not rest on forensics.' Kit loosened his cufflinks, little silver hooks that I only now noticed bore his father's initials. I couldn't imagine the Lachlan McCall I'd known wearing a suit. He tossed them like jacks from one palm to another. 'Could physical marks of a struggle prove he knew what he was doing or does she actually have to verbally say no?'

The next question was inevitable: did you actually hear her say no, Laura? I had to cut him off. I couldn't lie to his face.

'We've been through all this!' I said. 'Sex without consent is rape, Kit! End of!'

He recoiled in surprise. 'Yeah, I know that, but—'

The next question burst out of me at the same pitch. 'Do you believe her or not?'

I've reviewed this conversation endlessly, and what I think I really meant was: you do believe *me*, don't you? But Kit didn't know that, and he did what he always does when cornered: retreated into pedantry.

'You said yourself, she never actually spoke about it afterwards. It wasn't till the police turned up that she started talking, was it? So technically, I can't believe or disbelieve someone who hasn't actually said anything either way on the subject.'

He was right, of course, which must have made my next eruption all the more bewildering. 'I didn't know you were such a pompous, cynical *fuck*.' It was projection, of course; I was lashing out at him to appease my own guilt and confusion at my earlier performance. But I didn't understand that at the time. His face crumpled with the force of my accusation, but he stood his ground.

'I'm not cynical,' he said, with control. 'I'm . . . a *scientist*. You can't be emotional in a court of law. I'm just trying to see it like they do. I thought that talking through the mechanics of it might be helpful, that's how you process things. Not every debate has to be a personal attack on your values. You know your problem? Too much compassion.' He was shouting now. 'You can't go through life taking on other people's shit like this. You haven't got a filter.'

I'd started to cry halfway through his speech and I spat the next words.

'At least I've got feelings to filter out! At least I'm not a fucking robot.'

Now Kit looked like he was fighting tears. He tightened his fist around his father's cufflinks. 'That's not fair and you know it.'

We teetered on the brink of our first huge row until Kit, adhering to the habit of a quiet lifetime, acquiesced.

'Ok,' he said. 'I'm sorry, I shouldn't make fun. I can see how important it is to you. We just see things differently, that's all.' He kissed me on the forehead. 'We'll stay for the rest of the trial if that's what you want. Two days, though. I'm not taking any more time off university.'

'Thank you,' I said, noticing a prickling irritation that he'd rolled over so easily even though it was what I'd wanted. 'Just, I feel like I owe it to Beth, to go and see it happening on her behalf.' If he'd lost some of my respect, I'd tested his, too.

Kit folded up the map, getting the creases in all the right places where other people would wrestle with it. 'I think you've already done more than enough for Beth,' he said, tucking the map away into our suitcase. I stiffened but there was no side to his voice, no meaningful glance; it was just a turn of phrase that I had freighted with my own paranoia. 'Anyway, let's not keep going on about it,' he said, zipping up the bag. 'It's not like she's gonna *know*. Carol Kent told me she'd already left Cornwall. We'll never even see her again.'

17

KIT
18 March 2015

This cruise ship is built for old people. Handrails line every wall and half the chairs in the bar are those high-backed, high-sided chairs you see in retirement homes. I've found one that faces the room but hides my face so I can see but not be seen. I'd say at least half the passengers are pensioners, and – I didn't realise this when I booked – there are no children on the trip at all. I'm so used to public spaces being overrun with kids that the slow pace and muted hubbub here is unnerving. Or perhaps I'm just in the mood to be unnerved.

A cold beer is taking the edge off my earlier shock, although the woman I mistook for Beth is everywhere, snagging my vision whenever she goes to the bar. Worse, she keeps smiling at me and even though my rational mind knows it's not Beth, each time she does, there's a twentieth of a second where some primitive defence instinct turns my hands into fists.

'How's Louise's pregnancy going?' It's the first time Richard has mentioned it.

'Laura,' I say. 'She's good. Due in two months.'

Unable to resist, I show him the picture on my phone of the twenty-week scan of our twins, two curved full-moon skulls, the vertebrae like centipede feet. I still can't believe that the mono-chrome fuzz is my children.

'Double trouble,' says Richard after a cursory glance. 'You

took your time about it, though. I got Nadia pregnant on honey-moon.' The gaucheness gets my hackles up. No man who has ever had to masturbate into a cup in a clinic and who's had to watch his wife being injected and drugged, scraped and invaded, would ever say that. No man who has ever known his wife pushing him away because it's the wrong day, or watched her grimace her way through sex on the right one, would ever say that.

'One hit, bang,' he continues blithely. 'Almost too early if I'm honest, but it was the same with the second one, too. Up the duff, first month we tried.' I get it. Richard thinks *he* did it, that he can take the credit for the fluke of biology. It's a bit late, now that we're stuck together in a floating metal cell for four days, but it occurs to me I don't like Richard very much.

'I tell you what,' he says in what's clearly supposed to be a man-to-man voice. 'Get your oats while she's knocked up, because you won't be getting a shag for months now. You think the logistics of doing it with a seven-months-pregnant woman are tricky, that's nothing compared to having a baby conked out in bed between you.'

'Two babies,' I correct him automatically.

'Well, that's it, then,' says Richard. 'Your love life's over. Just wait until they're eighteen and then get into wife-swapping.'

I laugh too loudly to cover my grief. The truth is, I think I've already lost Laura in that way. The past few years haven't been like the beginning, obviously; we aren't immune to the law of diminishing returns. But right up until we started trying for children, I still wanted Laura and she still wanted me for the sake of it. The ghosts of our younger selves could always be coaxed out to play. That delight was underscored by a little more surprise every time, that despite all these years and all this time our bodies still fit together like they were made for it.

The first few months of trying for a baby are romantic; no, more than that, they're exciting: sex on the appointed day and, if that doesn't work, on the appointed hour. But sex by appointment soon became sex on demand, and once we went medical, something between us died. It was resurrected only briefly; there was a flurry of genuine desire I think around the beginning of the second trimester, but the mechanics were so challenging – that hurts, don't touch me there, maybe if I put a pillow under my hips, not like that, Jesus, Kit you're *squashing* them – that it didn't really count.

I can't tell Richard any of this. I can't tell anyone. I use my thumb to wipe the beads of condensation from my glass.

'It's all downhill from now on,' says Richard, setting his glass down on a frilly *Princess Celeste* beermat. 'Nothing puts you under pressure like kids.' At this, my anger turns to amusement, because after all Laura and I have been through, what harm can these longed-for, tried-for, paid-for and already beloved babies do? 'It'll be weird, having nippers after all this time with just the two of you,' he says. 'You've been together since you were kids, pretty much, haven't you?'

'Twenty-one,' I say.

Richard gives the corner of a smirk I've seen on countless other faces. I often meet people who almost pity me for settling down so early. Not just because I didn't get to play the field, but they wonder what I had of myself to contribute at such a young age. They don't understand that that's the whole point. I had the rest of my life to offer Laura. I look at these couples, who only met each other after their thirtieth birthdays, and wonder how deep a relationship like that can be. There is so much unshared life to intrude upon the marriage; so much opposing history. The defining event of my life is the defining event of Laura's. I don't know how couples who *haven't* been through something like that stick together. And if I've lost the carefree girl I first knew, then she's lost the boy with the brilliant future, and made him her present nonetheless. I know I'm not what she signed up for; after the meteor smash of the Lizard I was knocked off course and I never became who I was supposed to.

A slot machine nearby starts spewing out coins, the hard-rain clatter bringing me back into the room. The bar is emptying into the ballroom.

'We'd better go and get our name tags,' I say. 'The talk starts in a minute.' We both drain our glasses and set them back on the bar.

My knife is heavy in my pocket, a reminder of how close I came to a stupid, potentially fatal, mistake. The memory of the glinting steel raises a stark secondary dilemma. There is no point in getting Beth out of the way if Laura doesn't *know* she's safe. If I had to use my knife – or my hands, come to think of it, and I know how far I am prepared to go to protect my wife – would I tell Laura? Could I tell Laura? Could she live with me? Even knowing I only acted for her, would she still love me?

18

LAURA
10 May 2000

After the ordeal of the witness box, Wednesday was something of a relief. The professionals were being called in today. Our parts had been played, and nothing we could do now would influence the trial. The thought that I might have already sabotaged it was a steel band, tight around my skull, but at a base-rock level of morality, I felt no guilt. My lie was a route to the truth: therefore, truth it was. I thought back to Kit's mechanical performance in the dock, his dogged adherence to the facts, and knew he would never see it that way.

Kit and I were together in the public gallery for the first time, islanded by Jamie's supporters around us, our hands locked together. The day started with no preamble or recap, but rather recommenced as a play does after the interval. Polglase got to his feet and said, 'I'd like to call the officer in the case, Detective Sergeant Carol Kent, Devon and Cornwall Police.'

Kent took me by surprise when she swore on the Bible, and I didn't need to look at Kit to know he was rolling his eyes. I wondered for the first time about the assumptions jurors made about witnesses' choice of oath. If I worked in court, if it were my boring everyday job instead of a horrible, once-in-a-lifetime ordeal, I'd make a game of guessing who would swear on a holy book and who would not.

Kent's testimony filled in some of the gaps; we discovered that by the time the patrol car came to pick her up, Beth had recovered enough to give her name and her address, although it would be some hours before the whole story came out. I learned that Beth had met Jamie around a campfire the previous evening, and that she'd given him the brush-off then. We also heard about Jamie's subsequent arrest. Rather than the police-chase of our conjecture, he had approached a uniformed officer and calmly told him that he had been wrongly accused of sexual assault. That doesn't look good for Beth, I thought. But these were minor stops on the way to the prosecution's real concern; the physical evidence.

Much of it was boring information that seemed irrelevant to the case; Polglase spent half an hour mining Kent for details of Pip, the beautiful police dog I'd seen. As this went on, I realised that it is more tiring to be bored than engaged, and unable to see the point of this line of questioning at the time, I wondered if it was a trick to exhaust the jury. Kit's eyes kept returning to the digital clock on the clerk's desk, blocky red digits that tore their way through the seconds, their pace at odds with the dragging day.

There was a brief foray into physical theatre when DS Kent put on a pair of Thai fisherman's trousers. Polglase kept a straight face despite sniggers from the jury. 'DS Kent, if this were your garment, and you were entering into consensual sex, even partially clothed, enthusiastic, consensual sex, how would you allow yourself to be penetrated?'

'I'd undo them,' said DS Kent. 'There's physically no other way. You can't just slide them over your hips. If you wanted to access a woman's genital area, you'd have to be pretty forceful.'

I had a flashback to Beth's white leg, smeared with mud. I thought I'd suppressed my shiver but Kit closed his hand tighter on mine, as if to anchor me.

Polglase hooked his thumbs behind the lapel of his gown. (They must teach them to do this in law school.) 'Thank you, DS Kent. I believe that we are able to share with the jury the actual trousers worn by the complainant on the day of the rape.'

A junior officer produced a plastic evidence bag. Someone on the jury actually said, 'Oooh,' and was shushed by a frown from the judge.

'They're very muddied as you'd expect,' said Kent. 'There's fraying on the right-hand side where the tie is joined on to the body of the garment, consistent with a sharp tug.' She's on Beth's side, I thought; she wants a guilty verdict as much as I do.

'Thank you, DS Kent.'

Fiona Price swooped like an eagle. 'These are incredibly cheap trousers, a few pounds each, sold widely at festivals such as the one on the Lizard. Can you categorically state that this minuscule rip is not simply a result of poor manufacturing? Or that it's not wear and tear?'

Kent's pause was deliberate; a drawn-out beat in which the two women eyeballed each other. Kent blinked first. 'No,' she said, her voice sagging with resignation.

During the cross-examination, Price rebutted each of the prosecution's once-convincing arguments. It was like trying to keep track of a dozen shuttlecocks, all volleying back and forth at once. Surely the jurors must be dizzied by it; surely that was her intent. I had to close my eyes and will into my head the image of Jamie's face above Beth's, and I had to admit that the only reason I knew was because I *knew*.

Kit had been looking forward to the doctor's evidence. No, not looking forward; he still didn't want to be in court. Perhaps it is better to say that he was giving it a different kind of attention. I knew why; he was finally on firm ground. The trial had so far been little more than a war of words, but now things were to be presented in a way he could trust; observed through a microscope and coded as data. Now it was my turn to warn him not to get his hopes up.

'This is a consent case, it's not about identifying him,' I told him, while we were waiting for the judge to come back from lunch. 'He admits they had sex. The science is just background noise.'

Kit was about to reply but then the usher intoned 'All rise' and we leapt to our feet in anticipation.

'The prosecution would like to call Doctor Irene Okenedo,' said Polglase. Irene Okenedo stood about five feet tall, and the top four inches of that were hair, long braids twisted into a bun on top of her head. She looked about twelve years old, and I had the thought – acknowledged as ridiculous even as the mental image formed – that she should have worn her white coat, or a stethoscope around her neck, or something to convince us that she was a real doctor. I willed her to do a good job for Beth, even as I acknowledged with shame how incensed I would have been if she'd been biased towards the defence.

She swore on the Bible ('I never knew there were so many committed Christians in the professions,' muttered Kit sarcastically) and introduced herself as a locum A&E registrar, who had trained in treating and examining rape victims with the newly established Metropolitan Police Project Sapphire sexual assault unit. Her voice, low, considered, gave her the authority her stature denied.

'Thank you, Dr Okenedo,' said Mr Polglase. 'You attended Miss Taylor at Helston Police Station. How did you find her?'

Dr Okenedo cleared her throat. 'Physically, she was dehydrated and needed to eat, although she was overall healthy and well-nourished. She was extremely tired and dirty, with mud on her clothes and under her fingernails.' Now I remembered Beth digging into the earth. I examined my own clean fingernails and wondered how much of Kit was underneath them.

Polglase gave a slow nod, ostentatiously sympathetic. 'How about emotionally, psychologically?'

'I'd say she was in a state of post-traumatic shock. She was very withdrawn and monosyllabic, giving yes and no answers to my questions. She didn't want me to examine her.'

'Thank you Dr Okenedo. You've actually pre-empted part of my next question. The police asked you to examine the complainant for signs of recent sexual assault. Can you please report your findings?'

'Well, my first duty of care is to treat the victim as well as examine her. She was offered painkillers, and five milligrams of Diazepam, both of which she accepted. I began with an external examination, taking DNA from an oral swab, samples from under her fingernails and so on. The victim was withdrawn but co-operative. Her only prominent injuries were one bloodied knee, tiny cuts and grazes on both knees, and on the heels of her hands.'

I searched the jury for signs of my own discomfort here; only the Sikh man shook his head. Mumsy was alert, as though watching a particularly engrossing episode of her favourite soap opera. I wondered if she could remember being young. I wondered if she had daughters. 'One would expect, even in vigorous consensual sex, that a woman would shift her position to avoid bloodying her knees. Could these small injuries be the result of a woman held down against her will?'

'In my experience, yes.'

'Thank you, Dr Okenedo.' He shuffled his papers, surely for show. Beside him, Fiona Price's junior wrote something down with a squeaky pen, and the judge did the same. I burned to be able to read them.

'You did not arrive until four hours after the rape took place.'

'That's right.'

'We heard Miss Taylor describe how her underwear was violently yanked to one side prior to penetration. One would expect that this would leave some kind of welt on the skin in the immediate aftermath?'

'Yes.'

'With regards to the force used to pull at Miss Taylor's underwear, is it possible that any surface injuries caused by the tugging of fabric against skin would have faded by then?'

'Yes,' said the doctor, with the tiniest of shrugs. 'It's possible.'

'I'm sure. Four hours is a very long time in this context.' He looked down at his notes with a frown to emphasise how little pleasure it gave him to ask the next question. 'What about a genital examination?'

'I examined the skin around the vulva and anus, although this took some time as Miss Taylor found even lying down on the couch very distressing. When she was ready, I saw that no injuries were visible to the naked eye. But when I asked if I could attempt an internal examination she became further distressed, curling into a ball and repeating the word no.'

It was an echo, a virtual corroboration, of what I had said; my heart jumped for joy, then settled respectfully at the base of my throat.

'Is this response something you have seen before when examining someone in the hours immediately following an alleged rape?'

'Yes I have.' Dr Okenedo nodded gravely, and it wasn't until Kit gave me a weird look that I realised I was nodding along with her. I forced my head still, even though no one in the jury was looking my way. 'I'm sure you can understand that this is extremely traumatic and often physically painful for someone fresh from such an ordeal. We need a victim's consent before we can perform any kind of examination. She did, eventually, let me take a vulval swab, after a great deal of persuasion.'

I winced and pressed my knees together.

'Were you able to find traces of ejaculate on that swab?'

I might not have been there at the point of penetration but I'd been there at the point of panicked withdrawal; I pictured that incriminating silver thread as it snapped. Kit jerked upright and leaned forward, fully engaged now that the science was coming to the fore.

'I was,' said Dr Okenedo.

Polglase looked meaningfully at the jury, but even I could tell he was reaching. 'Can you think of a comparable sexual assault that you have attended, with comparable injuries?'

Dr Okenedo thought for a minute and then said a hesitant, 'Yes?'

He Said/She Said

'Can you let me know how long you spent on that?' Nathaniel Polglase was in his element now. 'Perhaps give me a minimum time it would take to do the procedure. Assuming that the complainant is well, and strong enough not to resist the examination.'

The word resist conjured stirrups and straitjackets. I couldn't fight the mental image of Beth, stripped to a hospital gown, knees forced apart for the second time that day. I felt my gorge rise.

'It can be done in about ninety minutes.'

'And how long were you with Miss Taylor, with all the additional cajoling and persuading involved in someone so traumatised?'

'I was in the police station for eight hours, from arrival to the final signing out. I'd say that I spent seven of those hours with the victim.'

'Thank you, Dr Okenedo. I have no further questions, Your Honour.'

Four hours for the doctor to arrive, plus an *eight-hour* examination; Kit and I would have been asleep in our tent while Beth was still being prodded by a stranger. I should have gone with her.

Fiona Price got to her elegant feet. 'We understand that the complainant *refused* an internal examination. Did you explain to the complainant why you wanted to examine her internally?'

'Yes.'

'And still she refused?'

'Yes.' Ask her why she refused, I thought. Let the doctor tell the jury the state Beth was in. But it didn't suit Price, so she didn't ask.

'So whilst we can't determine that there were no internal injuries, neither can we say that she did sustain any?'

'That's correct.' A single braid broke away from the coil of hair, and briefly bounced like an antenna before falling into the doctor's eyes.

'Dr Okenedo. How long have you been a doctor?'

'I qualified in 1997,' she said, tucking the braid behind her ear.

'Congratulations,' said Fiona Price, as though praising a Brownie Guide for her latest badge. 'You trained with the Met Police?'

'Yes.'

'Since moving to Cornwall, in how many sexual assault cases have you been the *sole* examining doctor?'

'Seven, including this one,' said the doctor, a ripple in her cheek.

'So, before you attended Miss Taylor, you had *only* overseen *six* rape cases?'

'Yes.'

Price went through some papers on her desk, letting the doctor's inexperience hang in the stuffy air. From a sheaf in her file, she drew a photograph, a beige blur from where we were sitting in the public gallery. 'These *tiny* grazes on the complainant's knees; could they simply have been pressure marks? The indentations made by a woman with the added weight of a man on top of her, during vigorous, *consensual* sex?'

'Yes,' said Dr Okenedo with reluctance. Fiona Price cut through the doubt. 'Did you find any bruising on the complainant's arms, for example? Anything to suggest that she had been held down against her will.'

That band around my head began to throb again, like someone was tightening the bolts.

'No. But again, freezing is a common response to rape. She might not have put up a fight.'

I studied the jury for signs of compassion. I got nothing, but they had barely had a chance to absorb the doctor's words before Price countered with, 'You're a doctor giving evidence about the injuries, aren't you?'

'Yes.'

'You're not a psychological expert on victim behaviour?'

'No.'

'So let's keep your evidence to your field of expertise, gained in your, what was it, three years of practice, shall we?'

How *dare* Price attack someone who spent her whole life help-ing victims? I knew that I could never do her job. 'To wrestle a woman to the ground in the first place, as Ms Taylor says happened, would still take considerable force, wouldn't it?' said Price.

'Yes,' replied the doctor.

'We have all seen Ms Taylor. She is physically fit and *feisty*, isn't she?' The mumsy woman on the jury was nodding in agreement; feistiness clearly wasn't on her list of attractive qualities. 'Even a healthy young man would not have been able to overpower her without a tussle. Did you find any bruising consistent with such a struggle?'

'No.'

'We have been told that the trousers were pulled aside with some force. The complainant wore nothing but a skimpy G-string underneath. My learned friend suggested that any welts caused by forceful tugging at these garments would have faded by the time you were able to examine Miss Taylor. But the kind of force Miss Taylor described would leave friction burns, would it not?

'Possibly?' My patience began to thin. Even I'd been more convincing than this, surely.

'A friction burn would not, in fact, vanish after four hours, would it?'

'No.' Irene Okenedo leaned forward on the witness box, but for support rather than in emphasis. Lean back, I willed her silently. Stand up straight.

'Were you able to find such marks on the complainant's body?'

'No.'

Nathaniel Polglase shook his head in derision. His perfect wig bounced precariously.

'You also examined my client physically,' continued Price. 'A genital swab showed the complainant's vaginal fluid. But that merely confirms what my client has maintained all along: that sex took place, doesn't it?'

'Yes.'

The barrister's voice had been slowly building in volume throughout this exchange. 'Did your physical examination of the complainant upturn *any forensic evidence* that she was forced into sex? There isn't a single physical injury, externally or internally, that rebuts the suggestion that there was nothing other than spontaneous sex?'

This time, she got the answer she wanted.

'No,' said Dr Okenedo. 'I can't say that.'

She looked like I felt; like she was watching a ball she'd kicked sail through her own goalposts.

'Thank you, Dr Okenedo. Your Honour, I have no further questions for this witness.'

As she sat down, Polglase rose, the movement smooth as two kids on a seesaw. 'Your Honour,' he said in a voice with the fight knocked out of it, 'That concludes the case for the prosecution.'

19

LAURA
11 May 2000

The day Jamie Balcombe went into the witness box, the sky above Truro was a clear, even blue. The defendant was borne in to the lobby on his usual flotilla before handing himself over to the dock officer.

The journalist with the bobbed hair was back in the atrium. When the court announcer called all parties in Crown versus Balcombe to Court One, Ali winked at her and whispered, '*Showtime.*'

We filed in to the public gallery, Kit and I taking care not to nudge or touch any of the Balcombe party. Someone was wearing an overpowering floral perfume that reignited yesterday's headache; my temples started to throb. The fiancée, Antonia, sat in the back row. Her outfit was calculatedly, almost parodically virginal, with a black velvet hairband and a frilly collar, little girl's clothes from a bygone time: she was only a pair of stripy stockings away from Tenniel's Alice. She craned to see Balcombe as he entered the dock. When he saw her sitting in the back, a hard, furious expression flashed over his face so quickly that I couldn't trust I had really seen it.

'Did you see that face he just pulled?' I asked Kit in a whisper. He looked at Jamie, who appeared solemn and respectful, and shrugged.

'He's always like that, isn't he?'

I watched as Jamie nodded sharply at the empty seats in the front row. Antonia stood up as though he'd yanked a string. Mouthing apologies at the dock, she scuttled forward until she was seated in the front row, where she sat, twisting her engagement ring nervously.

'You must have seen that,' I asked Kit, but he was fiddling with his watch, synchronising it with the court clock.

Jamie's walk from the dock to the witness box was ten, maybe twelve paces. He was overly polite and co-operative, loudly thanking the usher who held the door open for him and saying, 'Of course, of course, thank you,' as he was shown into the witness box.

'You may sit down, Mr Balcombe,' said the judge.

'Thank you, Your Honour,' said Jamie, bowing his head, 'But I prefer to stand.'

He was dressed in a different suit today. It was slightly oversized and made him look like a schoolboy on the first day of term, in a new blazer bought with growing room. As surreptitiously as I could, I looked sideways at the rest of them, immaculately dressed in the gallery next to me. It was hard to believe that Jamie's ill-fitting suit was anything but a device to disarm the jurors. His eyes looked bluer than they had before; was he wearing lenses?

He swore on the Bible. Of course he did.

Fiona Price smiled warmly at her client.

'Jamie, thank you,' she said, as though he had a choice, as though he was doing us all a favour by gracing us with his presence. 'Before we get to the night in question, I'd like you to tell me a bit about your education,' she continued.

'Thank you, Miss Price. I was educated at the Saxby Cathedral School, where I got four As at A level. I'm currently studying to be an architect. I got my Bachelor's in Architectural Design at Bath University.'

His voice was like expensive honey, dripping off a silver spoon.

'There are a further three years of study and work experience required before I'm eligible to join the Royal Institute of British Architects. I believe I'm right in saying that only veterinary surgeons train for longer than architects. After your BA, you spend a year working in the industry before returning to university for two years to complete a Diploma in Architecture, then more professional practice, more studying, more exams.' He gave a little smile. 'And after that you have to apply to RIBA and get a job. That's when the hard work *really* starts.'

A male juror in a blazer nodded as the tattooed man curled his lip.

'And how has this case affected your career?'

Jamie put a sag in his spine. 'It's destroyed it before it had a chance to get going, to be honest. I should be doing my year in industry now. I had a place with McPherson and Barr, the award-winning architectural firm. They were behind the new eco-estates being built on brownfield sites in the inner cities? It's a very prestigious placement; they only take on one graduate per year. I was just about to start my placement when these allegations were made and unfortunately the firm found it necessary to suspend me pending a verdict. So I'm currently rather in limbo.'

The story fell out whole. I tried to catch Jamie's eye. You can't lie to me, I thought. I know you. I've seen your rotten heart.

This was too much for Fiona Price; for a few moments she hung her head in grief at a young man's career derailed, before continuing. 'Your father is a successful housebuilder, is he not? In fact, it's not an exaggeration to say that he has an empire property, with new builds happening all the time. You don't need letters after your name to make your mark. Why not simply go into the family business?'

Jamie gave a bright blue blink. 'It's important for me not to cruise on my father's reputation. And besides, the future of the industry must be sustainable, responsible housing. That's where my interest lies. I suppose you'd call it a vocation as much as a profession.'

The irony struck me that a man on trial for rape could be so seductive.

'You are a young man with a very great deal to lose,' said Price. 'It must be very distressing, then, that your career is on hold.'

'Your Honour,' interrupted Polglase. 'My learned friend is seeking to try the matter on consequences, not on the evidence.'

Price didn't miss a beat. 'In order to assess how my client might act in a given situation, one has to take into account how much he'd have to lose if he did so. Let us move on, Jamie, to how this case has affected your personal life.'

'I've barely slept since they arrested me,' he replied. 'That's a long time to be tired. Even now I can barely believe this is happening to me.'

Fiona Price straightened a pen on her desk, then changed her tone.

'Are you married? Have children?'

'I'm engaged to be married. My fiancée Antonia is here.' He smiled at Antonia, who simpered in return. For the first time, it struck me; she had been in court since the first day of the trial, which means she couldn't be appearing as a witness. Why hadn't they called her? Why *wouldn't* they call her?

'Any children?' said Price.

'No children yet, I want to do things the right way round, but I'm very hopeful on that front.'

You could see the female members of the jury melting. Even I was struggling to superimpose the rapist's sneer over the blinking schoolboy in the dock before me.

'For the benefit of the jury, the alleged incident took place on the Thursday. Let me know how you came to be at the festival, and how you met the complainant.'

'Sure,' Jamie nodded. 'Of course. I was supposed to be going down to Cornwall with an old schoolfriend, Peter, but a couple of days before we were supposed to travel, he broke his leg abseiling with the Venture Scouts.' Of course he fucking did, I thought. What next? A digression into Peter and Jamie's other hobby of helping little old ladies across busy roads? 'So I went down on my own, on one of the coaches that was laid on from London, and then just set up camp.' He gave an embarrassed half-smile. 'It wasn't the best start to the festival. Pitching the tent on my own didn't come easily to me. Peter was the scout; that was his area really. Actually, once I'd got the tent all set up I felt a bit down about even being there. I'm not a natural loner. That's why I was walking around the camp- fires that night, looking for people to chat to.' He cast his eyes down, then up again, gave it the full Princess Diana. 'Lots of the campfire circles were quite big, it wasn't like you had to ask to join them. I didn't even look to see where I was sitting, I just sat down in the first space I saw. At first it was a general group chat, about other festivals they'd all been to, whether the weather was going to be clear for the eclipse, that sort of thing.'

'And how did you strike up conversation with Miss Taylor?'

He turned slightly so that I could see him in profile. I wondered if his looks would count for or against him.

'Well, once it got really dark, the guitars came out, and that meant we could only really talk to the people next to us. So we got chatting about this and that. She was really well travelled, she'd been to lots of festivals. I told her it was something I wanted to get into but my girlfriend wasn't really keen.'

Price twirled a pen between her fingers. 'So the complainant *knew* you were in a relationship?'

'Well, I thought . . . there was a bit of a spark between us.' Jamie looked apologetic, as if, even after what Beth had put him through, he didn't like to speak badly of her. The mumsy juror put her head to one side. 'She put her hand on my thigh when we were talking and I thought that I ought to get that out of the way, quick, so that flirting was off the table.'

Price needlessly straightened her wig. 'So we're talking about the previous night, the night before the eclipse?' Jamie nodded. 'Did you take any drugs at this campfire?' Kit and I looked at each other; this sounded like something the prosecution would say, surely. Polglase looked unsurprised.

Jamie looked at the public gallery and bit his lip.

'Jamie, you must answer the question. Did you take any drugs at this campfire?'

'I'm sorry. It's hard to answer this in front of my mum.' He sighed, long and deep. 'Yes. There was a joint being passed around and I took a puff on it, just to be sociable really. I thought it would be a way of being accepted as part of the crowd around the fire.' The tattooed juror nodded, as if to say, well, who wouldn't? 'I was just getting carried away with the atmosphere. But I coughed it out, I'm not used to it, and I said something along the lines of, this isn't my usual sort of thing. Beth took it from me and laughed and said, don't worry, it's a festival, normal rules don't apply.'

'And did Miss Taylor partake of this joint that was being passed around?'

'She did, yes, but only one go, I think.'

'Were you intoxicated?'

'I can't speak for Miss Taylor, but I certainly wasn't.'

I didn't realise I was drumming my fingers until Kit put his hand over mine to stop me.

'And how did this gathering end?'

'Well, at maybe midnight, I wasn't wearing a watch, she got up to leave. I asked her where her tent was and walked her back to it. I wanted to make sure she got there safely. I stayed there till she'd zipped herself in.'

Staying positive during this performance was like trying to hold water in my fist.

Price leaned forward on her knuckles. 'And you weren't tempted to make a move on her there? In the dark, when it was just the two of you?'

Jamie palmed his eyes and waited a long time before answering. 'Yes, I was tempted, ok? We'd hit it off. But I didn't do anything about it.'

I looked at Antonia. Her face was a mask. What the hell is going through your mind, I wondered.

'So, knowing that she was vulnerable, and with both of you under the influence of cannabis, you left her to sleep on her own?'

'Of *course* I did,' said Jamie, as though anything else was unthinkable.

'Of course. And now we come to the morning of the alleged incident itself. How did your day begin?'

'He deserves a fucking Oscar,' I muttered to Kit. He took my hand and smoothed it flat from its fist, but like everyone else he was mesmerised by Jamie.

'The next morning was the day of the eclipse,' said Jamie. 'I thought I'd walk by Beth's tent to see if she wanted to watch it with me. She wasn't in, but I bumped into her leaving the main field, and she said she was going to find somewhere to watch it in peace. It was pretty full-on in the main field with all the music and shouting, and I didn't fancy it either, so I said I'd go with her.'

'And what was her response?'

'She didn't object.' Jamie was emphatic. 'I wouldn't have gone otherwise. So, we walked a sort of winding way around until she suddenly stopped in this field full of circus equipment. I thought, hang on, because it wasn't the best vantage point, there were loads of lorries and stuff in the way, but then I realised: what it did have going for it was privacy. She made a little space on the grass between two caravans and I sat next to her.'

'Was this by mutual agreement?' Price leaned forward; the jurors leaned forward; everyone on the public gallery and press benches seemed to lean forward, as though Jamie was a vortex, pulling us all in.

He handled his words with kid gloves. 'I would say *unspoken* mutual agreement,' he said. 'Maybe that was naive of me. If I'd known what was going to happen . . . but it was all so spur of the moment. We just looked at the sky; well, at the clouds. Nothing happened for a while and then it all sped up. We only saw the sky for a second, but the way it went dark was just uncanny. It's an incredible thing to share with someone, incredibly intimate. You've got all these people around you, but you feel like it's just you and the other person and there's this darkness coming out of nowhere.' *No*, I thought irrationally; this man is not allowed to be sensitive to the beauty and power of an eclipse. I won't allow it. Jamie cleared his throat. 'And then, when it started getting light again, I said – what I meant to say was that it was amazing or incredible but there was a sort of Freudian slip and what I actually said was, wasn't that romantic. And I knew it wasn't a particularly appropriate way to describe it, but it was.' He looked at Antonia helplessly and his voice lowered in respect to her. Her face remained inscrutable but her right hand repeatedly twisted her engagement ring. I didn't know whether to feel contempt or pity. 'And the mood just took us over, and after we started kissing it just went from there really quickly. It was so spontaneous I couldn't even tell you who instigated it.'

Price put out a hand to stop him. 'This is important, Mr Balcombe. The kiss was mutual? She never pushed you away, she never asked you to stop?'

'Absolutely not. Absolutely not. If I'd read the situation wrong, I would have stopped in a heartbeat.'

'We come now to the mechanics of sex,' said Miss Price. 'Could you please clear up my learned friend's confusion about Miss Taylor's trousers, and how they were removed?'

'She loosened them rather than removed them,' said Jamie. 'They came open along one side; I wouldn't have known how to start. And she tilted her hips up towards me . . .' he put his head briefly in his hands, to show what an ordeal this was for him. His dad nodded his encouragement. 'And I pulled her knickers to one side, and that was it. It wasn't rough, I wasn't hurting her, it was – I'm so sorry, Antonia.' Jamie took a few seconds to gather himself. 'It was just . . . exciting sex. I've never done anything like it before or since. And then this couple, or that girl, came blundering in and the next thing I knew, I was being told I was a rapist.' His voice rose up here, and Price shot him a warning look.

'The witnesses, Miss Langrishe and Mr McCall.' All eyes swivelled towards us. All eyes except Antonia's. The diamond on her finger flashed with each rotation. 'And how did you respond?'

'Well, I was more trying to get Beth to say what had really happened. But she just went mute. I mean, I know it was awkward, I wasn't exactly pleased about being caught in the act either. You've got to understand how belligerent they were. They, *she*, just decided what I'd done.' He looked at me and I hoped the sweat that broke out on my face wasn't visible to the jury. 'It was surreal, it was like a farce. I even thought they were joking for a second, and then when they said they were going to the police I realised this was really happening, and they were serious. I never thought for a *second* that I'd come to court. It's a bad dream.'

'Thank you, thank you.' Price squared off a sheaf of papers for punctuation. 'Now we come to another moment. After the initial accusation was levelled at you, you left the scene. Why, if you were innocent?'

Jamie Balcombe sighed deeply. 'I had a bit of cannabis in my pocket. I'd got someone to roll me a joint and bought it.'

This was news to me, and evidently to the prosecution too; Polglase passed a note to Carol Kent and a little flare of hope rose in my chest.

'And you didn't think to tell the others this?' said Fiona Price. 'After all, you were hardly at a church fete, were you?'

Jamie spread his palms wide. 'I had some woman I'd never even seen before accusing me of rape! It was hardly a relaxed atmosphere! My thought process was, I'll just get to a point where I could dump the drugs, because I don't want to be charged with possession. If I'd just raped someone, do you really think I'd be worried about a little joint in my pocket? But that witness, he was right behind me.' He nodded towards Kit, who sat up straight in his seat. 'He'd have been able to see anything I threw out. So when I saw this big crowd all going back to the main stage, I let myself get swept up in them, and then I thought, I know, there's that little police station set up near the entrance. I'll go there, and say there's this woman accusing me of something awful, and then this'll all be sorted out. I was sure she would've calmed down by then anyway.'

'How long did it take you after the alleged incident to hand yourself into the police?'

'About an hour, I think. Well, after I'd thrown the joint away, I had a bit of time to think about a bigger issue,' said Jamie. 'I mean, I *did not rape* the complainant. I didn't, I couldn't. But that doesn't mean I hadn't done anything wrong. I'd cheated on my girlfriend, my fiancée. It was a moment of madness but not in the way he's making out.' He nodded at Polglase. 'I wavered a bit because I thought, if this becomes official, even if all I do is get ruled out, the chances of Antonia finding out what I've done go up massively. So I went back to my tent for about half an hour and thought it through, but then I realised I had to do the right thing, and put an end to the misunderstanding. I never in a million years thought I'd be charged with rape.'

'What reason might the complainant have to accuse you of doing so?'

Jamie swallowed hard. 'I can only think it's embarrassment. After all, it wasn't the most dignified thing.' He shuddered, like it pained him to be so unchivalrous. God, he was good. 'But I think it's clear from her performance in the witness box that she has convinced herself she's telling the truth. That's the worst part of it; she's a victim, but not of me. I had hoped that by now she would have got the help she so obviously needs.'

20

KIT
18 March 2015

The *Princess Celeste*'s ballroom has all the atmosphere of a strip-lit shopping centre. At the opening health-and-safety talk, a rep from the tour company has a passenger list, ticking off each name when the corresponding badge is collected. By the time Richard and I get there, all but a handful of the badges have been given out.

The first lecturer takes the stage, and I identify the source of the disembodied voice I'd heard over the tannoy. Professor Jeff Drake lectured me for three years at Oxford. He will film and narrate the eclipse as it happens, and perform what he calls a 'post-mortem' on the return journey. I've got one ear on him as I creep to the back of the ballroom, where the passenger list is left unattended on the welcome table; it takes me seconds to scroll down and eliminate Beth. Then, mindful of Laura's new identity theory, I find a vantage point halfway down the ballroom and methodically scan every female – not as time-consuming as it sounds; men outnumber women by about two to one here, and young women are scarcer still – until I can be sure that they aren't her. I have been as thorough as I can.

Beth is not on the ship.

There's still a chance she's on land. If I *were* going to look for her, find her before she can find me, my greatest chance would be in that bar. During the shadow, there will be thousands of tourists scattered across dozens of hills. We won't even know our own vantage point until the morning of the eclipse, when the scouts go out to assess the cloud cover on the island. But for the next twenty hours, until we dock in Tórshavn Harbour, there will be no confrontation with my past.

I think I deserve another drink.

I catch Jeff Drake at the bar. He remembers me at once; 'Christopher!' he says (thanks to my name tag, my full name is already beginning to prick my ears the way *Kit* does at home). His voice kicks me back through the corridors of time and into his rooms overlooking the Isis. I half expect to look down and see the battered Adidas Gazelles I used to wear.

'I often wonder what became of you,' he says, when we have finished telling each other what a small world it is. 'Académe's loss must have been industry's gain.' It is a long time since I felt shame when I told anyone what I do for a living now, but when I do, he is gracious and interested; if anything, he is disappointingly undisappointed in my failure to complete my doctorate. But Jeff is as famous for his diplomacy as he is for his intellect. When our conversation is interrupted by an elderly Canadian lady earnestly asking the difference between a star and a planet, he speaks to her with the same respect he showed me when I was his student. My wasted potential is a stone in my throat, and I wash it down with red wine.

Richard has fallen into conversation with a party of astronomers from Wales and it's soon established that there are some serious eclipse chasers here. Tansy, a vast, ruddy woman about my mum's age, threatens to show us her lucky underwear. 'Never been clouded out with my lucky knickers on!' she says, topping up my glass.

'Still got it,' I tell Richard under my breath. As notes of eclipses *passim* are compared, a pecking order naturally establishes itself. Richard takes a register of names and eclipses viewed on the back of a tour itinerary. Our undisputed master is a nonogenarian Californian man who has seen nineteen eclipses, but if you divide the number of eclipses by our ages (and I can't resist the calculation) then I have seen by far the most eclipses relative to my years.

Wine just keeps appearing. The temperature in the bar rises and I peel off my Faroese sweater to reveal my Chile '91 T-shirt underneath. The effect on Tansy is like Clark Kent ripping off his suit to reveal the Superman outfit beneath. '1991! You must have been a babe in arms.'

More wine, and I realise I have stopped noticing the ship's keel, which means either we are motionless on a millpond sea, or the lurching feeling in my brain is in perfect synch with the vessel's slow rocking. Richard challenges everyone within earshot to break the world record for the most eclipse chasers ever in one photograph. As an eclipse virgin he is not eligible to take part, but he herds us into his viewfinder. There's a *Princess Celeste* souvenir baseball cap lying on a table and, without asking whose it is, I pull it down over my eyes and stand at the back. I'm a black peak and a ginger beard. When he's got his shot, Richard runs the end of a straw up and down his list, his lips moving as he does the mental arithmetic. 'This picture,' he says brandishing the image on his phone, 'represents a total of 103 discrete total-eclipse-viewing experiences.'

'If that's not a practical use of mathematics I don't know what is,' I say, to laughter all round. Mac's been telling me I'm not funny my whole life, and even Laura doesn't get all my jokes, but tonight I realise I've been looking for appreciation in all the wrong places. Here, they love me. Everyone wants to hear my eclipse stories. I can talk freely about them for the first time since before Cornwall, because these people are so fluent in the technicalities of the experience that I can leave out everything else – the interruptions and aftermaths that would have soured even perfect conditions – and they will *still* be interested. My new friends hang on my every word. Tansy inches closer. I look at the wine glass in my hand only to find I've moved on to rum. Someone produces a selfie stick and there are more pictures, and I keep the cap on. I'm introduced to another man in a Chile '91 T-shirt; naturally we embrace like long-lost brothers. I note with satisfaction that while his strains over a proud beer belly, mine still hangs more or less straight from my shoulders to my belt. No wonder Tansy finds me so irresistible. Conversations with my new friends repeat, circle and blur. I believe there is some singing and at one point I tell an amazing story, I think.

'Maybe you should calm down a bit,' says Richard, gently taking my glass from my hand and setting it down.

I am indignant. 'Me and Tansy are just friends.'

'What are you on about? Your bunk's two feet away from mine. I don't want to be woken up by a face full of your puke.'

I'm too drunk to stop drinking but still sober enough to know that fresh air is my friend. On the top deck, I thud on to a recliner and then I have a quick power nap. When I wake up, someone has covered me with a blanket. Below and around me, waves shatter against the sides of the boat. The wind has blown the clouds to one side and the moon is a thick waning crescent. There are more stars than blackness above us; multi-armed galaxies swirl before my naked eye and meteors are as frequent as London buses. I haven't seen a firmament like this since Zambia. Laura's absence is the only flaw in a perfect evening. I cannot remember the last time I felt so at home.

21

LAURA
11 May 2000

'If *I* were wrongly accused of a crime as heinous as rape, I'd stand my ground, but you ran away, didn't you?' said Nathaniel Polglase.

I held my breath as Jamie evenly returned his gaze. 'As I said, I was just going to get rid of the drugs in my pocket.'

'No. You ran away because you had been caught in the act of raping the complainant Miss Taylor, and you hoped to get away with it, didn't you?'

'No.'

'Do you remember when you first approached the police, in their Portakabin cell?'

Jamie couldn't hide a panicked glance at his counsel. Fiona Price's nod was almost invisible. I wondered if Kit had seen it. I wondered if the jury had seen it.

'Yes?'

Polglase looked solemn. 'The jury has heard DS Kent describe Pip the drugs dog, but I will recap for their benefit as well as yours, Mr Balcombe. There were two police dogs on site at all times that weekend, including a four-year-old Alsatian called Pip, an incredibly sensitive, highly trained animal. To him, even residual traces of drugs are as obvious as a lit joint would be to the human eye. Where was Pip when you were questioned, Mr Balcombe?'

'He was next to me,' said Jamie. Kit and I sat up straighter. 'He was in the corner of the room with his handler.'

Polglase pressed on. In cross-examination he finally seemed to be enjoying himself.

'Did Pip go for you?'

Jamie's face expressed exactly my feelings when I'd been cornered by Price.

'You don't remember Pip, because he didn't go for you, did he? He did not smell the cannabis you claim you were so eager to get rid of.'

'It was gone by then. And I'd had it in a little plastic bag.'

'In those circumstances traces would still be detectable to a finely trained nose. Mr Balcombe!' Polglase unveiled a rich tenor. Everyone in the courtroom rose an inch from their seats. 'This joint never existed in the first place, did it?'

'It *did!*' I heard the little-boy whinge he'd used on Beth in the aftermath.

'It's a smokescreen, concocted after the event, something you thought up in the journey between the festival and Helston Police Station, isn't it?'

'No!'

'How much did you have to smoke around the campfire?' Jamie didn't say anything. 'Your statement is here if you need it,' prompted Polglase. I knew from my own stint in the witness box that when counsel hold out your statement, they are handing you the rope for your own noose. I was glad now of the experience; it helped me understand Jamie's present discomfort.

'One puff,' he admitted.

'And did you like it?'

'Not really.'

'Why, then, did you go and buy some?'

Jamie cast his eyes to the ceiling, as though the right answer would be written there. The jurors wouldn't have looked out of place with oversized soft drinks and cartons of popcorn.

'You're a cunning young man, aren't you?' said Polglase.

'What?'

'Isn't it far better we condemn you for the lesser crime of possession than focus on your true transgression, the brutal rape of a vulnerable, lone female?'

Jamie just shook his head. Along the bench from me, his mother mirrored the action. The press were scribbling furiously in their notebooks. I had no need to record anything; even now, I can virtually recite some of the key speeches from the case. Memory, I was discovering, acts completely differently when you give it a little notice. When you know at the beginning of a day that every word will count, even extraneous details imprint themselves in your long-term consciousness. It's when events sneak up on you that things become disordered. There should be different words for the different ways we can remember.

'You were lying on one of those counts. Which one?'

'I'm not lying,' said Jamie. Oh, but you are, I thought to myself, and they're going to see through you, just like I did.

'It almost doesn't matter. Let's press on,' said Polglase. 'I refer you to Ms Taylor's testimony from earlier. It's not just your opinions of the rape that are in conflict. You both arrived on the same day, which was the Wednesday, but that's about all you agree on. Let's go through the day before the rape, shall we? It was you who first touched the complainant Miss Taylor's thigh, wasn't it?'

Jamie appeared back on safe ground. 'She pressed up against me.'

Antonia stopped twisting her engagement ring; the diamond caught a shaft slanting through the skylight, and beamed a point of white light across the room.

'Whose suggestion was it that you accompanied her back to her tent?'

'Mine. I thought she was vulnerable.'

'In fact, you insisted on accompanying her despite her several assurances that she would prefer you not to, isn't that so?'

'No.'

'Her actual words at the threshold of the tent were, I believe, "Not in a million light years." Polglase paused for a pantomime wince. '*Ouch*,' he said with a little moue. 'No one likes to be rejected, but that's got to sting, hasn't it?

'No,' said Jamie. 'Because it didn't happen.'

'It's humiliating. You wanted to teach her a lesson, didn't you?'

'That's not true,' said Jamie, but the charm had peeled away from his voice in what I hoped was a precursor to losing his temper. Let them see the real you, I thought. Let them see what I saw.

'Ms Taylor says that she stayed awake until she was sure that you had gone, that she was frightened to go to sleep while you were there. Is that the experience of a woman who would the following day enter into consensual sex with you?'

'She's lying because she was embarrassed at being caught.' Jamie's tone was somewhere between patience and pity. 'Maybe she hadn't done anything like that before, I don't know. It was out of character for me, too. It was a spur-of-the-moment thing. The whole *sky* was changing.'

Polglase turned to the jury. 'Mr Balcombe now appears to be excusing his actions on the position of the planets. Whatever next? Star signs?' The tattooed juror stifled a laugh. 'This isn't out of character for you, though, is it, Mr Balcombe?'

'That's because I didn't do it.'

'You have a history of pursuing what you feel entitled to, regardless of the cost to others, don't you?'

He drawled the words, like they were too boring to bother with. 'I really don't know what you're on about.'

'Were you successful on your first application to the McPherson Barr internship?'

Kit and I exchanged puzzled glances. 'No,' said Jamie, but he no longer sounded bored.

'So how did you change their minds?'

'I made approaches to them at a charity event, and persuaded them how keen I was.'

I could just picture him making his way through the drinks and the canapés, gliding on his entitlement like a set of castors.

'But what about the original, successful applicant?'

Now Jamie scowled. Kit raised his eyebrows at me. I nodded an acknowledgement, but couldn't take my eyes off the witness box.

'What exactly did you tell Octavia Barr, the CEO of the company? I remind you that you are under oath.'

'I told her that the other applicant had had a caution for affray. I didn't think that was the kind of person who deserved such a responsible position.' Jamie tilted his chin squarely at Polglase, evidently to underline how seriously he took his responsibilities as a citizen.

'So you denigrated the successful applicant and usurped him?'

'You make it sound far more underhand than it was.' Jamie's colour didn't rise and his skin remained matte. There was none of my nervous sweating or of Dr Okenedo's crumbling under pressure.

'I make it sound exactly like it was,' said Polglase. 'You don't take no for an answer, do you?'

'That's not true.'

'Isn't it true that you have a history of responding to rejection with spite?'

'It's not the same thing at all – is he allowed to do this?' Jamie asked his barrister.

Nathaniel Polglase addressed his reply to the judge. 'It's relevant to character, Your Honour.'

'Go on, Mr Polglase,' said the judge. Fiona Price waited until the judge wasn't looking to shake her head.

'Isn't it the case that you harassed her at the campfire, you followed her back to her tent and you followed her on the morning of the eclipse, even when she asked you to stop?

'No.'

'Far from mutually deciding to observe the eclipse from the caravan park, you followed her, biding your time until she was at last in a location so remote that you were able to take, without permission, the thing you had wanted since you first set eyes on Ms Taylor. Didn't you, Mr Balcombe?'

'*No!*' The word was a whip cracked through the stuffy air. Polglase gave us a few moments to absorb its resonance, taking a sip of water. Jamie did the same.

'Mr Balcombe. Did Ms Taylor actively say, I would like to watch the eclipse with you, or words to that effect?'

'No, but not *everything* has to be said, does it?' He spoke at half his previous volume.

'I would argue that it does, Mr Balcombe. I would argue that the very basis of this case is about spoken, unequivocal consent, but despite repeated use of the word no . . .' Polglase paused to let the word reverberate around the court and I held my breath: if he attributed this to eyewitness statement, Kit would know it could only have come from me – and it wasn't something I would have kept from him. 'You exercised its very opposite, didn't you?'

I was spared, again, for now. I looked to Jamie.

'*No,*' he said, closing his eyes.

'You raped Miss Taylor, didn't you?'

Jamie's eyes snapped open and looked straight at Polglase. 'NO!'

No. No. No. No. The word vibrated on repeat until Polglase picked up the echo's tail.

'No further questions, Your Honour.'

The abruptness of his closure took everyone by surprise. Jamie Balcombe gawped at his parents as his chance to respond was snatched away. Sally Balcombe made a soothing, shushing sound he couldn't have heard.

Fiona Price had only one further witness to call for the defence. Christobel Chase was a rangy girl in a bright green dress who took the secular oath in a loud, clear voice and confirmed that she had met Jamie Balcombe in her first week as an undergraduate at the University of Bath, in her capacity as the steady girlfriend of Jamie's old rowing team-mate.

'I'd like you to talk me through an incident that occurred in the final term of your first year at Bath, when the defendant Mr Balcombe came to your assistance,' said Miss Price.

'Of course,' said Christobel. 'So, it was the Christmas ball and I was really the worse for wear. I must have had about two bottles of wine over the course of maybe eight or nine hours.' She glanced at Antonia, then at Jamie in the dock. 'At some point I wandered away from the students' union and ended up staggering around the campus. I couldn't find my way home. Jamie noticed I was missing from the table and came out to look for me. I could barely stand, but he walked me back to my room, held my hair while I vomited into the toilet – not my finest moment – and helped me into bed and put a bucket by my side in case I got sick again in the night.'

Price nodded. 'How would you describe his conduct?'

'He was a perfect gentleman,' said Christobel. 'I'd want him looking after me every time.'

'Thank you, Miss Chase.'

Nathaniel Polglase scratched underneath his wig.

'Miss Chase; you were at the time already seeing one of Mr Balcombe's close friends, is that right?'

'Laurence, that's right, yes. We're still together.'

'Had he taken advantage of you then, there would doubtless have been repercussions within his social circle?'

'Well – yes. But he wouldn't have.'

'There is a world of difference, members of the jury, between homing in on a young woman alone, miles from home, at a remote festival, and treating with care a woman with whom one is already socially involved, and, may I say, for whom one already has respect. One doesn't soil one's own doorstep, as I believe the popular saying goes.'

'Mr Polglase!' said the judge. 'This is veering dangerously close to a speech. Do you have a question for the witness?'

'Apologies, Your Honour.' Polglase closed the file in front of him with a click. The gesture was probably supposed to hide his frustration but only underlined it. 'That concludes my cross-examination.'

Judge Frenchay peered at the clock. 'We'll save closing speeches, and my summing up, for tomorrow,' he said. I had expected this, but still felt the internal plunge of disappointment; we wouldn't be there to see it. Our tickets back to London were booked for the morning, and we barely had money to eat, let alone pay for new train tickets and another night's accommodation.

As we stood for the judge's exit I thought hard. Tomorrow's speeches would probably make further reference to my own testimony. Fiona Price would try to discredit me again, drawing attention to the inconsistency in my statements. Truro Crown Court was the last place in the world I needed to be.

'It's scary how convincing he was,' I said on the way down the hill. Kit took my hand and circled his thumb on my palm.

'What if he did it but he *thinks* he's innocent?' he asked. 'What if she thought it was rape but he didn't?' He wasn't playing devil's advocate, more acting on some kind of atavistic masculine solidarity. It was the first time that I realised that men must fear the accusation just as women fear the attack. 'I mean, is it even possible?'

We came to a stop at the bridge. The Kenwyn flowed fast beneath us, two washed-out plastic bags chasing each other downstream.

'Honestly?' I stared into the water. The plastic bags parted ways at a fork in the current and disappeared under the bridge. 'I think some men hate women so deeply, they don't even know it. They don't think in terms of violating consent because they don't think about consent, full stop. Maybe Jamie Balcombe's one of them. I don't know.'

'That's not a defence though, is it?' asked Kit.

'Jesus, no,' I said. 'That makes it even worse.'

The following Tuesday, we let ourselves into the flat to find the light on the answering machine flashing red.

'This is a message for Christopher McCall and Laura Langrishe,' said a familiar, officious voice. 'It's DS Carol Kent from Devon and Cornwall Police here, it's just gone three o'clock on Tuesday the sixteenth of May. We've got a verdict. Took the best part of a day, but the jury found Jamie Balcombe guilty by a majority of ten to one. The judge decided to sentence the same day, what with people travelling from so far and wide. He gave five years, which is the maximum penalty he could have awarded for an assault at this level. The judge made a bit of an example of him. You might like to know that Elizabeth was in court for the verdict and I think she's very happy with it. I know we are.' There was a brief pause. 'Anyway, I wanted to thank you for your co-operation in the case. Your evidence was a big part of what put him away.'

The light flashed off and there were no more new messages. The jury saw through him, I thought, and it was as though someone had undone a giant bolt in my shoulders, screwed so tight for so long that my arms felt momentarily dislocated. I collapsed against Kit. 'It's over,' I said to his thumping chest.

Second Contact

22

LAURA
19 March 2015

The sun inches slowly across the moon, stealing light as it goes. Kit is on a deserted mountainside, eyes glued skyward, Beth sneaking up on him under cover of the shadow. *This isn't over yet*, she says again and again. There's a jagged shard of glass clutched in her fist but it's my palms that are bleeding. I'm screaming Kit's name in warning but nothing's coming out; Jamie Balcombe's hand is tight over my mouth.

The nightmare wakes me but the vivid little film takes a full minute to dissipate. I cradle my belly and roll on to one side, eyes wide. A streetlamp throws grubby London light through the slats in the blinds and the bedside clock tells me it's 3.59 a.m., about the same time I woke up yesterday. I heave onto all fours and reach for Kit's pillow, focusing on the cool cotton under my palms. But the mental picture is burned into me, like a retina scarred from staring at the sun. I text him.

Are you awake?

What I mean is, are you alive?

I regret it as soon as I've done it. It's one thing to be a nervous, controlling shrew (my words, not his – although he must think it) around the house, but to nag him across timezones is inexcusable.

I'm losing my grip again. When he doesn't answer within sixty seconds, my heartbeat becomes audible. One of the babies turns a somersault inside me, creating the horrible sensation of a roller-coaster's plummet and, just like that, I'm in freefall. When anxiety wins, the mad part of me peels away from the rational one. My capable self is standing on a distant shore, watching in horror as I flail in rushing currents of my own making. That's how it feels now, as I call Kit's phone. It goes straight to voicemail three times in a row. A new image comes to me; Kit leaning over the ship's railings, caught unawares as Beth first knocks his phone overboard and then flexes her hands to push . . .

Next thing I know, I'm shivering in my dressing gown and Uggs in my dark living room, my bump resting on my crossed legs. When I fire up the iPad my face is reflected in its surface for a second, a hollow-eyed ghoul with long white hair and concave cheeks. Night dissolves the day's discipline and the double *oo* of the Google logo returns my unblinking stare.

Thought blurs into action. I'll call the ship itself, ask the staff to check Kit's cabin, or maybe, in case that seems mad (ha!), I could just ask whether everyone on board the *Princess Celeste* is accounted for. But the only number I can find is for the tour operator, and it turns out they're not answering their phone at five minutes past four in the morning.

I look for FAROES ECLIPSE STABBED TOURIST CRAZED WOMAN DIES KILLED CHRISTOPHER MCCALL BETH TAYLOR PRINCESS CELESTE NORTH SEA on all the newswires: Press Assocation, Reuters, BBC, Sky News; surely, if something's happened, they will have it covered between them. Momentary peace when my search draws a blank is immediately surpassed by the prospect that something terrible has happened, it just hasn't been reported yet.

I run the same words through Google just in case. *Stop it*, says my rational mind. *You're making yourself ill. You're flooding your babies with stress hormones.* The dominant part of my brain sticks two fingers up in response, and I hit return. This time the internet sends me to YouTube and instinctively I watch through my fingers, like a child. I can tell even through the filter of my hands that none of this handful of clips is *the video*. The footage is amateur but it's recent. One is from this afternoon, a ten-minute film of the sun setting over the North Sea. Tentatively, I click. There's no music, just a little film someone's made of the sinking sun. I watch the whole thing, focusing on nothing but the shimmering of bronze light on silver water. Wave by wave and breath by breath, I slowly return to calm. Repeatedly calling Kit's phone now seems at best embarrassing, at worst dangerous. He'll assume that there's been a medical emergency. I pick my phone back up.

> *Sorry ignore me I'm fine just had a bad dream. Babies etc all good.*

When the sunset video ends, the queue of videos in the sidebar shifts. I'm braced for a freezeframe from *the video* inching its way up the screen, but it seems I'm safe. These films are posted by scientists, not hedonists. Eclipse chasers who post online tend to divide into two tribes: the serious amateur astronomers Kit's with now, and the new-age rave contingent. The former hugely outnumber the latter, so if you wanted to access *the video* you'd probably need to include the word festival in the search.

There's only one more clip with *Princess Celeste* in the caption.

> *Eclipse-chasing cruise. Drunk guy rapping on the* Princess Celeste *HILARIOUS*.

The clock in the corner of the screen tells me it's twenty minutes past four. I can tell that sleep is hours away. I could do with a bit of light relief. I highlight the video and click play.

23

LAURA
16 May 2000

Our walk-up on Clapham Common Southside was on the fourth storey, with a little balcony overlooking the green. In winter, bare branches left visible the mansions on the north side, but in summer the world beyond stopped at the nearest treetops. To get to our flat, we had to climb eighty-five stairs that zig-zagged tightly between windowless landings. The other three flats were accessed via the back alleyway, so once the street door was closed, we were home. There were no neighbours to tidy the stairs for; nothing to stop us nipping down to pick up the post with no clothes on. We used to joke about getting a fireman's pole installed to shave seconds off the morning commute, a ritual we still spoke of with the self-conscious, self-important tones of the very young and newly professional. The wallpaper in the stairwell was ancient and peeling, here and there flaking back so that you could see a layer of vivid green paint that Kit said was probably Victorian, contaminated with arsenic. I thought he was joking until I threatened to lick it once and he pulled me back in panic.

There were only two real rooms: our bedroom at the back, and a long, sloping living room-cum-kitchen. The appliances were older than we were, the bedroom door didn't shut properly, the Vent-Axia fan in the shower was so loud that you couldn't listen to the radio while someone was in the bathroom. There weren't enough power points to accommodate all of Kit's technology and black wires writhed in nests at every socket and underneath the telephone.

The space was tiny – half our stuff was in boxes in Adele's loft – but it was ours. It was *us*.

Kit's eclipse map had pride of place above an old futon that had once belonged to Adele. Once or twice a week, we'd convert it from a sofa into a bed for when Mac would roll out of some sewer of a club south of the river or – more and more frequently – when Ling had locked him out. I got used to finding him passed out in our sitting room, a bucket for vomit at his side. He was already missing a molar, and his face was starting to sink in on itself, presaging how Kit might look at fifty. He was only twenty-two.

Kit had persuaded our local newsagent to order in the West Country newspapers. He needn't have; the Balcombe verdict made the nationals. He came home from the newsagent with a copy of *The Cornishman* under one arm and half a dozen tabloids and broadsheets under the other. We read the lot on our balcony with a pot of tea and a pile of toast between us, in a silence that was nervous only on my part. Kit started with the regional papers. The *Cornish Times* carried an interview with the farmer Rory Polzeath, who was not only financially ruined but also emotionally devastated by what had happened on his land.

I blackened my fingers flicking manically through the headlines, wondering which, if any, might give my lie away. ECLIPSE RAPIST GUILTY said the *Sun*. PUBLIC SCHOOLBOY PROTESTS INNOCENCE said the *Daily Mail*, in a report that concentrated on the price of Jamie Balcombe's family home, the cost of his education and his father's tenuous connection with the Prince of Wales. There was nothing about my testimony, so although my instinct was to throw the paper across the room in disgust, I handed it to Kit to buy myself a few more minutes.

A female columnist in *The Times* claimed the rape was an example of all that was wrong with the 'alternative scene'; it was effectively a warning shot to parents whose children went to music festivals. I felt myself clench all over. The *Telegraph* took it even further. The headline JUDGE MAKES EXAMPLE OF FESTIVAL RAPIST presided over a lament on Jamie's ruined career.

'Whose side are they fucking *on*?' I said to Kit, forgetting for a moment my own vested interest in the reports. 'Are we meant to feel *sorry* for him?'

'Oh, dear,' he said, adding the *Telegraph* to his reading pile.

In the *Daily Mirror*, I zoomed in on one phrase that made my blood run cold.

> *There were some tense moments in the trial, including, at one point, a crucial eyewitness who wavered in her testimony.*

I didn't dare read on, not with Kit close enough to see over my shoulder. I stuffed all of the *Mirror* in between the pages of a discarded sports section I knew neither us would ever read, then pushed both into the recycling bin.

The bobbed journalist, the one I'd pegged as a *Guardian* writer, actually turned up in *The Independent*. Georgie Becker's byline photograph was a good decade out of date but her words were the first to resonate with me.

THE MOST COMMON REACTION TO RAPE IS FREEZE:
AT LAST, ONE JURY RECOGNISES THIS
AT LAST, ONE JUDGE SPEAKS UP FOR VICTIMS

> 'The sound of his flies being unzipped, it was like hearing the barrel of a gun click into place; you're a hostage, you just do what you're told. I couldn't fight him off, I froze. I just wanted it over with.' So ran the testimony of the twenty-year-old victim of Jamie Balcombe, the man last week convicted of a prolonged rape during a festival to mark the total solar eclipse in Cornwall in August of

last year. Her powerful words shatter the accepted narrative that to resist rape is for the woman to fight back until she is red in tooth and claw. For most rape victims, fear is paralytic.

Rape cases in which the parties are acquainted are notoriously difficult to prosecute. So what went right, here? The Balcombe case is highly unusual in that there was corroborative evidence – in a court of law, a woman's word alone is not enough – and Mr Balcombe's victim admitted that she might not have gone to the police had the rape not been interrupted by two passers-by, whose testimonies were crucial to the conviction.

At this point I froze myself, but the incriminating detail was left out. I was overcome first with relief, and then disgust at this relief when Beth's ordeal had been so much greater than mine. I read the rest through incipient tears.

Balcombe's victim showed courage in reporting her crime and sticking to her story in the face of a blistering cross-examination by the defending counsel Fiona Price. She was made to relive her sexual history, recounting all partners from her loss of virginity up to the day of the event. Throughout this gruelling ordeal, her rapist was repeatedly referred to as one more 'sexual partner'. She was repeatedly told that she had had a 'moment of madness' in the eclipse that she was ashamed of, and that it was her own lack of self-control that was on trial here. At one point she screamed, 'This is worse than being raped.'

The judge, Mr Justice Frenchay, twice stepped in and ordered a break for the victim to recover, but this is a symptom of his sympathy; Miss Price's interrogation was entirely legal, and is all too common a feature of rape trials where consent, rather than identity, is under scrutiny. That the victim's sexual history was a benign catalogue of serial monogamy is, must be, irrelevant; the point is that the recounting was tantamount to psychological warfare, an attempt to erode what dignity the victim had preserved.

Change is coming: when the so-called Rape Shield Law, the amendment to Section 41 of the Youth Justice and Criminal Evidence Act, comes into force later this year, it will greatly curtail the questioning of a complainant about her sexual history and behaviour.

Passing sentence, Mr Justice Frenchay said: 'You are an arrogant opportunist who deliberately preyed upon a lone woman, and, when she rejected you, isolated her further until she was defenceless against your attack. You did not respect her pleas for you to stop. You did so knowingly and with a total disregard for her physical or emotional wellbeing. It is your sense of entitlement, and your perception of women as mere commodities, that causes me to give you the maximum sentence for this crime.'

Justice Frenchay and the jury have made a small step towards lasting change for all victims of rape. Let us hope we are finally on the threshold of something significant.

'What happened to the *Mirror*?' Kit shuffled beside me. 'I'm sure I picked one up.'

'Haven't seen it,' I said, deliberately not looking at the recycling box.

'Weird.' He scratched his neck and opened the *Express*. 'Oh, it's our friend from the court.' Ali, or Alison Larch, to give her her full byline, had secured her interview with Antonia Tranter after all. It was a double-page splash. The redhead gazed plaintively up at me from beneath the headline, JAMIE, I'LL WAIT FOR YOU. I let it fall to the floor.

'I know how you feel,' said Kit. 'I've read enough for now.' He stacked the papers in a neat pile. 'Remember I'm meeting Mac for a lunchtime pint.'

'Are you sure that's a good idea?'

'If you can't beat 'em, join 'em. At least this way I know what he's doing. That's if he turns up.' Kit spoke with the air of an old soldier going into battle.

The only sounds in the flat were sirens and horns, faint from the street below. It was the first time I'd been alone in days and suddenly my own company was not enough. With Kit out with Mac, and Ling welded to her baby, there was no one I could call for a casual game of pool, or to go mooching around Brixton market.

To busy myself, I heated milk for coffee and read through Alison Larch's *Express* piece. Antonia, I read, had forgiven Jamie his 'infidelity' – 'What the fuck?' I shouted at the page – but was still reeling from what she called a miscarriage of justice. The Balcombe family and their legal team had been so sure the case would be thrown out of court and, later, that he would be found not guilty, that they had even bought a flat for him to move into on his release, in which Antonia 'rattled around' on her own. Now poor Jamie was in the same wing as serial rapists and child molesters. Beth, of course, wasn't named, but was referred to as Jamie's accuser rather than his victim throughout. I was fighting the urge to claw the paper to shreds when an unknown number called my mobile phone. I flicked the kettle on, then picked up.

'It's Beth.'

My silence was only shock; I wasn't used to reading about people in the national press and then having them call me. And I had assumed that, because the trial had gone her way, we wouldn't hear from her again. Beth took it the wrong way.

'Elizabeth Taylor?' I think it's only hindsight that gives her words a petulant tone. 'From, you know, from Cornwall?'

'Sorry, yes, of course I know who you are,' I said. 'God, it's good to hear from you. How are you? What's the right thing to say? Congratulations doesn't seem appropriate.'

'I don't know,' she said. 'Relief is the main emotion right now. I'm at home in Gedling. I don't know what to do with myself to be honest.' It took a few seconds for me to realise that she was asking my advice. I had none to give her; I was the very *last* person she should look to for wisdom. 'Anyway, I know you said to only get in touch if it all went to shit, but I couldn't let you go without saying thank you. Not just for testifying in court but for taking over just after it had happened. I don't know, if you hadn't come along, how it might have gone. I probably wouldn't have had the balls to do this on my own. And the CPS said your testimony made all the difference. The way you fought my corner. So, you know. I owe you one.'

I looked again at the picture of Antonia Tranter in the *Express* and hoped Beth hadn't seen it.

'Not at all,' I said. 'Anyone would have done the same.'

'No, they wouldn't.' Beth's voice was urgent, excited. I know I'm not imagining the fervency that overtook her then. 'I don't think you *know* how special you are. And I need you to know it. Do you believe in karma?'

'I'd like to,' I said carefully.

'Well, I do, and I reckon you're due a charmed few years.' I could hear the smile in her voice. Behind me, the kettle started to roar and the microwave pinged. 'I'll let you go, you're probably busy. I just wanted to say, thanks.'

'You're welcome.'

I finished the piece in the *Express*, drank my coffee and thought about Beth. In another lifetime, I thought, had we met in different circumstances, we might have become good friends.

24

KIT
19 March 2015

I lean on the port deck of the *Princess Celeste* and look out over Tórshavn harbour, fighting the kind of hangover I haven't known since I was a student. The smell of fish on the breeze turns my stomach. I'm in my pyjama trousers under my orange coat and my feet in my boots are bare. I haven't brushed my teeth, washed my face or turned on my phone since last night. Richard has gone to get me a kill-or-cure coffee. The air here is so clean it feels like you drink it rather than breathe it, and I need every cubic inch of oxygen I can get to blast away last night.

Our ship casts a shadow that reaches right across the bay. From here you can take in the whole town, clustered on the water's edge, the surrounding mountains rolling into the vanishing point. The basalt cliffs erupt from the sea like the volcanoes they once were. Red-painted houses dot the grey townscape. They look like Lego, and the crowds that line the streets, in primary-coloured walking clothes, seem to have escaped from a toybox. We've got a day to kill before the eclipse and nothing booked until this afternoon, when we'll take a minibus up into the mountains. The thought of it summons bile to my throat.

I haven't got the energy to find my way to the breakfast room, let alone go hunting for the stone-clad bar.

'I want to go back to bed,' I tell Richard when he arrives, my caffeinated elixir of life hot and bitter in a styrofoam cup.

'No chance,' he says. 'We're going exploring.'

The coffee hits the spot. In the cabin's cramped en-suite, I brush my teeth, shower, dress and feel almost human. Only then do I check my phone. There's a welcome message from Føroya Tele, wishing me a happy stay on the islands, followed by some unsettling texts and then three missed calls from Laura.

I'm awake now. I reply. *Heavy night. Glad you're ok tho. Call you later. Xxx*

I leave it behind; I'll catch up with her later. Richard and I fill our wallets with smooth new ten-krónur notes and hit downtown Tórshavn. On the streets, entrepreneurial locals are peddling the usual novelty tat: T-shirts, baseball caps and viewing goggles. I buy the lot, in duplicate this time, so my children won't have to fight over who gets what; I am determined that, unlike me and Mac, they will both be eclipse chasers. Our Faroese jumpers are on sale everywhere and I'm not sure Richard and I look as cool as I thought we did. Every shop and café is full to bursting.

Richard has bought a paper map. 'These wooden warehouses used to belong to the Danish royal family,' he says as we get up close to one of the red houses.

'Let's get this party *started!*'

'All right, sarky bollocks,' says Richard, then nods to a nearby bar. 'Hair of the dog?'

We each buy a pint of pale, hoppy lager that's been brewed specially to commemorate the non-event. Sólarbjór was apparently brewed in complete darkness and under a full moon. It could have been made by elves and pixies under a magic rainbow, I still wouldn't be able to finish it. I take a couple of sips then push it away. I'm never drinking again.

In the National Museum, there's an account of the last total eclipse over the Faroes, in 1954. The crowd jostling to read about the eclipse is four deep and it's not very often you get to see that. Richard reads a quotation from the then state geologist. '"Rain and foggy, impossible to work",' he says. 'Well. At least it's not raining.'

'Not yet,' says a woman in a rain hood. 'Fifty per cent chance of precipitation tomorrow according to my app.'

I belch coffee into my fist.

'How you feeling now, mate?' says Richard.

'I would say, really profoundly depressed.'

'At least you've seen one before,' he says, and then dramatically flings his hand across his eyes. 'Please, oh gods of the weather, don't make me die a virgin!'

'Ha.' I'm warming to him after all.

My phone's ringing when we get back into the cabin. Laura's picture fills the screen and there's a notification in the corner telling me I've had, oh hell, sixteen missed calls from her. My hand trembles as I swipe to answer.

'What's happened?' I say.

'Have you fucking *seen* yourself on the internet?' she says shrilly. Acid swirls in my belly. She must have gone through my computer forensically to find my social media accounts. It's completely out of character, but she's not herself lately. I should never have left her alone.

'It's all under an assumed name.' I can hear the squeak in my voice. 'I didn't even put my face on it, I covered my tracks.' I absolutely did; my avatar is my Chile '91 T-shirt, I've even disabled the locations on the few pictures I've posted. There's no way she can have—

'What are you talking about?' she says. 'The whole fucking thing is your face, in close-up! You've gone and brought her nicely up to date, haven't you? Making a complete spectacle of yourself on the internet with a fucking *rap*.'

'A rap?' I echo, but the word unlocks a memory. Last night; I mucked around in front of a camera; I – oh *God* – I made up a rhyme, didn't I, listing all the eclipses I'd seen. I can't remember anyone saying it was going on social media; but I can't remember telling anyone not to share it either. Hello guilt, my old friend.

'You shouldn't be allowed to drink,' she shouts. I hold the phone away from my ear, like men do in sitcoms. Yes, I'm in trouble now, but she's not talking about my Facebook account and this will in time be written off as a mistake, not a deception.

'You need to get them to take it down, now,' she's saying when I put the phone back to my ear.

'Look, I'm sorry, I had no idea anyone was going to upload it.'

'It's rule number one! Don't get yourself pictured. You got yourself fucking *filmed*.' She's crying now.

'Please don't get wound up, it's not good for the bab—'

'Don't you tell me what's good and what's not good for the fucking babies!' I can picture in my mind's eye the veins that stand out on her neck when she loses her temper.

'I'm sorry,' I repeat, because I haven't got a leg to stand on. 'But listen, Laura, she's not here. I've scoured the whole ship. That's why I had a drink, to celebrate her not following us.'

'Well she doesn't need to be on the ship now, does she? She knows where you're going. She's got the *name* of the ship! She's got the name of the town! She'll be on a plane now. If she's not already there.'

We've never had a big argument over the phone before. I search for the right, calming words. 'All the flights to the Faroes were booked months ago.'

'No.' She speaks with conviction. 'There's one chartered flight going from Inverness this evening that had, as of yesterday, two seats on it but that's literally the only way to get there. And one of them's already gone this morning, since you decided to declare your whereabouts.'

Anger takes the edge off my guilt. I've fucked up, but Laura hasn't kept her side of the bargain either. She promised not to wind herself up, looking for trouble on the internet. She didn't just stumble upon this while she was shopping for groceries or sending a work email. I can picture her sitting on the sofa, legs crossed under her bump, rising hysteria with every clickthrough. But I don't say anything; I never do.

'The chances of her being the one who booked that ticket are *tiny*.'

'Yeah, well, the chances of her finding us in Zambia were pretty fucking remote, and she managed it then.'

'That was different. We were at a festival, she was shooting fish in a barrel.'

There's a crackly silence that lasts so long I have to check my phone for reception. Five full bars are lined up like tubular bells. She's sulking. I don't know how to do this long-distance. I wish I could touch her.

'Laura, you're overplaying the risks.'

'But if she does find you?'

Then I'll do whatever it takes to deal with her, I think, but I know that will only set Laura off again. 'I tell you what,' I say. 'Let me have a look at the video first. I'll call you back later, I promise.'

Back on deck, reassurances to Laura of my continued anonymity ring hollow as a fellow passenger, someone I don't recognise, high-fives me and congratulates me on my 'poem'.

Shit.

In the lobby I cocoon myself in my favourite chair, plug my earbuds into my phone and, with an uneasy churning deep inside, I go onto YouTube. It can only be *Eclipse-chasing cruise. Drunk guy rapping on the* Princess Celeste *HILARIOUS*. Silently vowing not to touch anything stronger than lemonade for the rest of the cruise, I tap to play. I didn't think a person could blush with no one watching, but my cheeks fire as I look at my purple drunken face dribble and slur its way through a list – I would call it a monologue, rather than a rap or poem, but there you go – of all the eclipses I have witnessed since Chile '91. If, in my drunken idiocy, I believed the

Erin Kelly

cap an effective disguise, I was mistaken; it had by that point ridden up on to the top of my head, not even casting a shadow on my face. There are two saving graces, I suppose. The first is that my name tag was on my jumper, which I'd long cast off by the time of the performance, the second that my name isn't mentioned at all, not even Christopher, let alone Kit. I'm introduced as 'a fine gentleman and scholar'. The second is that my only concession to Cornwall '99 is 'Clouded out/twist and shout', to boos all round. I'm not sure whether they're sympathising for my bad meteorological luck or the terrible rhyme. I don't, in fact, think it's quite the security breach of Laura's imagination, but there's no denying I'm a *massive* dickhead.

A hand on my shoulder makes me jump in my seat.

'You found Darren's vlog, then?' Richard drops into the chair opposite me. 'He just told me about it. I was going to curl up and watch it now.' I don't know what my face does but Richard softens a little. 'Ah, come on, you weren't that pissed. It's funny, it's charming.'

'I fucking hate technology,' I say, head in my hands. 'How is this even allowed?'

'Cruise policy.' Richard now looks sheepish. 'I meant to tell you, didn't I. When you got up to go for a slash?'

I think back to the front of the evening; he must mean when I was sneaking out to check the passenger list for Beth's name. 'They gave out stickers to anyone who objected to being filmed because so many camera crews were going to be around that they couldn't get release forms for them all. Standard practice. You have to tell people, about photographs, that social media is part of it all.'

'This is bullshit!' I say.

'It sets a dodgy precedent,' he concurs. 'That's the way it's going though. Opt-out culture. Really, we should have opened up a debate about it at the time but no one said anything, so . . . Why, are you worried about what they'll say at work?'

Work! I hadn't even thought about work. 'Yes,' I say. 'Work.'

I eventually locate Darren the vlogger, a hirsute man in a tie-dy top attacking the fruit machines in the little arcade. He grins in recognition but motions me to stand to one side until he finishes his game. He jabs at chunky flashing buttons for what seems like forever and it's all I can do not to turn it off at the plug.

'What can I do you for?' he says. 'Enjoying your fame?'

I hope he's not going to make me beg. 'I'm really not,' I say. 'I need it taken down now. This could have serious repercussions for me at work.'

'You said it was ok at the time,' says Darren, but even he doesn't look convinced.

'I was drunk as a lord. I'd have said yes to anything.'

'Fair dos,' he says. 'I'll just have one more game here, see if I can get my stake back.' He jingles coins in his pocket.

'Can you do it now?' I've never been any good at giving orders to other men, and I'm braced for the *fuck off*, but Darren re-pockets his coins.

'Wait here,' he says. 'I'll go and get my tablet. I'll do it in front of you.'

In the arcade, primary-coloured lights flash at me in the dark, calling my hangover back into being. I step into the lobby before I have some kind of seizure. Darren's back in three minutes. His eyes are downcast in what looks like shame, but he's biting back a smile.

'I can take it down no problem, I'm doing it now.' With a few deft swipes he's deleted it from his account. I can tell by his face there's a 'but' coming.

'What?' I say.

'It's a bit awkward. You went low-level viral. Professor Brian Cox retweeted it. Nineteen thousand hits in the last two hours alone. It's gone right to the top of the YouTube sidebar.' Darren's trying to look apologetic but can't disguise his glee. I turn on my heel, eyes down; the pattern on the carpet seems to writhe under my feet. Anyone who's looking for an eclipse chaser will see it. I might as well have given out my co-ordinates. My pulse quickens as the reality of seeing Beth again gets one step closer, and it's only partly in fear. Maybe I need this showdown. Maybe my drunken subconscious made a decision my sober self hasn't had the balls to make for fifteen years.

25

LAURA
18 May 2000

I *meant* to confide in Ling about my week in court. The confessional urge wriggled inside me like something alive on the train to her new place in Green Lanes. I changed lines at Leicester Square, and made the journey underground, with no real sense of navigation. North Londoners still complain that South London is a lawless outpost because there's no Tube network but that was its beauty for me; you could come to *know* it. London, south of the river, seemed a true city, a sprawling blend of communities linked by buses and overland trains. North London to me then was a patchwork of islanded villages, reached only by Tube and never joined up above ground; circles on a map, separate as stars. Turnpike Lane, for example, I thought, as I counted down the identical Victorian streets to Ling's new flat. Where the fuck *was* this place? I couldn't put it in context. Pre-gentrification, the Harringay Ladder was the kind of place you don't go unless you've got a bloody good reason, the kind of place South Londoners have never even heard of.

Four months later, this would be its attraction.

Ling opened the door to the basement flat while I was still on the uppermost step. Baby Juno was sprawled face down on her shoulder. 'He's done it again,' she said, bursting into snotty tears. 'He's fucked off with my cashcard and I haven't seen him for forty-eight hours. I can't do this on my own, Laura. I can't fucking do it. Sorry, come in, come in.' I stepped gingerly over the threshold into chaos. When people talk about being knee-deep in nappies they mean it rhetorically but Ling was literally wading through piles of babygros, vests, blankets and cloths, the laundered tangled in with the filthy. While I stood wondering where to sit, a clothes horse in the corner of the room collapsed under the weight of tiny clothes and the story on my tongue slid back down my throat. I couldn't dump my shit on Ling when she was in this state. I think even then I knew that that was it; I could feel the secret setting solid inside me. If revenge is a dish best served cold, confession should be dished up piping hot or not at all.

'I think I'm dying,' said Ling. 'I think I'm actually dying of all this, it's going to kill me.' She swept her hand around the room at the mess – it lingered for a second on an empty vodka bottle next to the changing mat – then her fist settled in a gentle full stop on Juno's little back. Kit and I had known for a while that Mac's benders were more than a new father letting off steam, but this was the first time I understood that Ling's not-coping ran deeper than the usual baby blues. They were hurtling towards their crises in parallel with us, just one reason why they still don't know the whole of it.

'Hey,' I said, taking Juno from her. 'Hey, it's fine.' The baby's hair was a wavy black corona around a golden orb, a total eclipse in negative. She was my niece in all but blood and, I now think, one reason Kit and I were in no rush to try for children of our own; not only that loving her stemmed the parental instinct in us, but the way Mac and Ling imploded after her birth. We valued our relationship too much to let that happen to us.

I sent Ling to bed, and while Juno kicked on her playmat I made a half-arsed attempt to clean up the house, putting a wash on and folding clothes, bagging up the newborn things Juno had already outgrown. When she cried, I gave her her bottle, burped and changed her. As Juno melted into sleep on my chest, I analysed her upturned face for her parents' features. It was hard to tell whose eyes she had when they were closed; she had Mac and Kit's pointed nose. Once she hiccupped and blew a milk bubble that stayed unpopped on rosebud lips that were all Ling's. I let my own breathing fall in line with hers: three breaths of mine for each she gave. Into her tiny little ear, I whispered my confession. She is still the only person I've ever told. I would be grateful later that Ling had fallen asleep. She might have understood what I did in court, but then she would also have been party to my later doubts, and with her as my witness, I would have had to act.

The Friday after the verdict, I left work to find Yusuf, the monolithic security guard, blocking someone from the building. I didn't give it much thought; the office where I was temping was in that strange place between the City and Soho, close to the British Museum and the old Virgin Megastore, and it wasn't unusual for tourists and shoppers to ask to use our toilet. I sidestepped Yusuf only to feel a warm, dry hand grip my forearm.

'There she is. Laura!'

I could see why Yusuf hadn't let Beth in. She had on purple flared jeans, a yellow belly top and half a dozen necklaces. Her hair was tied up in a messy topknot. I felt caught red-handed in my temp's uniform of dress and heels. Beth seemed so much more *herself* than I was; no, she seemed more like *me* than I did.

'Have you heard?' she said. I noticed her eyes were bloodshot. 'They didn't ring you?' I put up a hand to show Yusuf I'd got this, and led Beth by the arm to the street.

'Who didn't ring me?' I said.

'The police.'

My legs went boneless as my mind leaped to the only conclusion; somehow they had found proof of my lie. Had word of my testimony got back to Beth? Had she given me away? She wouldn't jeopardise her own rapist's conviction on a technicality. Would she?

'No.' I thought for the first time what a tidy, contained word 'no' is; hardly any room for the speaker's voice to shake and betray. It's a gift to the liar.

'Jamie Balcombe's been given leave to appeal. Carol Kent rang me this morning. She didn't want me to find out from the papers.' Beth sniffed loudly, like a child. 'Jamie threw out his old legal team and the new one reckon they've got some new evidence that might get the whole thing back in court again, and I can't, I *can't* live through that again.'

Neither could I. I knew already that I could not lie convincingly twice; spontaneity had made me credible in a way that premeditation never would. I would have time to think about it now and get nervous. Reverting to my police statement would probably mean acquittal for Jamie Balcombe and a perjury conviction for me.

'*Shit*.' I said. Beth misinterpreted my self-serving panic as concern.

'Shit's about right,' she said. She blew her nose with a dirty tissue. 'Sorry. You could probably do without this. Only, I can't talk to my parents about this, they've been through enough, and my friends don't get it, and you were *there*, Laura.' A colleague emerged through the revolving door of my office and wished me a good evening. 'Sorry,' said Beth. 'You've probably got somewhere to go.'

As if anything was more important than getting to the bottom of this.

'No,' I said. 'I think we should . . . I think we should go and get a drink.'

We walked in silence, heading into Holborn. My thoughts were a tangled ball of string. Ideally – for her peace of mind as well as mine – Beth had blocked out much of the attack, and took my word for the truth. Or the details of my statement in the witness box had somehow escaped her, which didn't seem likely. It was possible, though, that she knew perfectly well that she hadn't said no, or hadn't said no in my hearing, and understood that I had put my neck on the line for her. Either way, she had as much to lose as I did. The tangle of string in my mind's eye pulled itself into a tight knot that could only be undone with scissors.

We ended up in a tatty old man's pub just off New Oxford Street, one of those mysterious central London pubs that never actually seems to have anyone in it yet remains there year after year.

'Bottle of white?' I said, when we'd climbed rickety stairs to the second-floor bar.

'Thanks,' said Beth. 'But what'll you have?' The joke was a toe in the water: is it ok to be frivolous, given how we met? My answering laugh seemed to reassure her. The wine was only Echo Falls white zinfandel but the barman put it in an iced bucket, which I carried rather self-consciously to our table.

'So . . .' I filled Beth's glass first, even though my mouth was watering for something cold and sharp to dull the edges. 'What's happened? What, *exactly*, has happened?'

'Jamie's got a new legal team,' she said. 'Obviously you could tell at court his family had money, but I didn't realise how wealthy his dad was until I read the fucking papers.' She'd seen the reports, then. That line in the *Mirror* – *a crucial eyewitness who wavered in her testimony* – couldn't have escaped her. I braced myself for the challenge, but it didn't come. 'That man, Jim Balcombe, his wallet's a bottomless pit. He can basically afford to keep going till they get the result they want. I looked up the new lawyers. They specialise in getting men like him off the hook.'

I took a large acid gulp of wine. 'But what new evidence have they *got*?'

She frowned into her glass. 'They reckon they've traced someone who was at the campfire, the night before the eclipse.'

So this had nothing to do with my testimony. Relief elbowed fear to the side; concern for Beth rushed in.

'What can they say, if they weren't even there?' I asked. My glass was empty already.

'Apparently they saw us cosying up together. It backs up his argument that we'd been flirting beforehand. Which is bullshit, by the way. I mean, I *had* met him the night before, that's how come he was tagging along on the eclipse, but I spent the whole night trying to get away from the slimy fucker.'

I remembered this from Jamie's cross-examination. It was one of the few points where his temper had come close to the boil. 'This could be good, though,' I said. 'I mean, a new witness might come down on your side. If they were actually witnessing harassment, maybe that could come out under cross-examination.'

'Yeah, right. Jim Balcombe's probably already written them a cheque,' said Beth. 'Even if he hasn't, some poor junior barrister versus their QC?'

'It worked last time,' I said. Beth was dismissive.

'I promise you, they're not going to stop until they buy the result they want.' She sloshed the wine around her glass. It slid down the inside of the bowl like olive oil. 'I'm trying not to get too angry at his family. I mean, I'm sure my parents would do the same if it was me, if they had the cash. They all think he's innocent.'

'Well, he didn't convince a jury,' I said.

'Thanks to you.' I couldn't work out whether her smile was one of gratitude or conspiracy, and I hopped away from the quicksand subject of my evidence.

'What happens now?' I said.

Beth refilled my glass, then turned the bottle upside-down in the ice bucket. 'As I understand it, it's only *leave* to appeal. It's like the step before you actually get to do it, and even then there's no guarantee that it'll go back to court. So there's a lot of steps to go through.' Her voice wobbled. 'The worst-case scenario for me is that the judge grants him the right to appeal, and that's successful, and it goes to a retrial and this time he gets off.'

'It won't happen,' I said, more to convince myself than her.

Beth caught her silent tears at the point of exit, pressing a napkin to the inside corners of her eyes. I motioned for the barman to bring us another bottle. 'D'you know, this is the first time I've been anywhere on my own since the Lizard?' she said unsteadily. 'I've tried to go out with my mates, tried to act like nothing had happened, but I couldn't get further than the garden path before I lost it. They're losing patience with me.'

'Do they know what happened?'

'A couple of people know I was raped, but I haven't told them I'm the Jamie Balcombe girl. I think one of them's guessed, though; she kept asking where my trial was. You know, it was a high-profile case, it happened at the same time, it doesn't take Sherlock Holmes. If I'd said Cornwall she'd have known, so I didn't. Or maybe she doesn't; maybe she was just showing interest and it's my paranoia. I wish I'd never told anyone, to be honest. Funnily enough, I wasn't nervous about coming here to see you. It's . . .' Beth brightened, then her face clouded just as quickly. She dropped her chin to her chest. 'No, it sounds ridiculous.'

'Go on,' I said.

'You feel safe.' It seemed to me then that it was the scale of the honesty required rather than evasiveness that kept her eyes on the table in front of her. 'I don't feel anything bad could happen if I was with you. I know it's daft. But you did save me, after all.'

I went quiet while the barman refilled our glasses, then set the bottle back in the bucket.

'I think I should've done more,' I said, quietly. 'I should've come with you, afterwards.'

'That's kind,' she said. 'But there's not a lot more you could've done. You'd only have been hanging around outside the cells.'

'The *cells*?' I was stunned. 'They didn't take you to hospital?'

'They've got a special room at the station but they were using it to interview someone,' she said. Her shrug acknowledged the horror of her treatment but also the time that had passed since. 'The cells were the only place I could get any privacy while the police were waiting for a doctor to turn up.'

'Oh, *Beth*. As if being examined wasn't bad enough.'

'Pretty grim,' she said, recrossing her legs. 'But the thing that sticks in my mind now is how it was afterwards. They gave me a pair of tracksuit bottoms to go back in, but they kept all my clothes from the waist down and they didn't have any knickers for me. They drove me back to the festival to get my stuff from the tent – I just took my clothes and left the tent where it was – but the whole time, I didn't have any underwear on. I was so aware of it; I felt like everyone knew.'

She screwed up her face against the tears and I understood that, actually, I would lie a second time for her. They could try Jamie Balcombe again and again, and I would lie every time to get him put away.

We drank in silence for a while, and the uncomfortable silence that always follows premature intimacy wedged itself between us. The frost was broken only when I followed her gaze to the empty pool table.

'I don't suppose you play?' she asked, in much the same way someone might ask whether you had a helicopter licence or the prime minister's telephone number.

I grinned. 'I'll wipe the floor with you.'

Beth got change at the bar and stacked the twenty-pence pieces on the side, then flipped the topmost one.

'Call it,' she said, her hand over the coin.

I blew chalk dust off the cue. 'Heads.'

'It's tails.' Beth broke, scattering reds and yellows evenly across the baize. She was shorter than me, and took on one tiptoe shots I could make with both feet flat on the ground, and between shots she walked the perimeter of the table, viewing it from every angle.

'Where are you living now?' she said, like she was catching up with an old friend.

'Clapham Common.' I bounced a red off the bumper and into the far pocket. 'Little top-floor flat.'

'I had to move back in with my parents for a bit. Until I get used to being on my own again.'

'And how's that going?' I loved my dad, but living in his house and reverting to his rules seemed intolerable.

'I dunno, they *mean* well. But I didn't have much choice. I had to stop working, so I can't pay my own rent.'

Was there no area of her life he hadn't ruined? 'What did you do, before?'

'I've never had a career, as such.' I couldn't work out whether her glance at my clothes was admiring or pitying. 'After college I worked as an au pair all over Europe. Before Cornwall I was working in a bar, just while I was deciding what to do when I grew up,' she smiled ruefully. 'I tried to go back afterwards but I couldn't hack it. Moving through the crowds like that. Bodies everywhere.' She shrank in on herself. 'You forget how much *bigger* they are. You don't realise they're built differently to us, how strong they are.'

I leaned the cue upright, ready to offer a hug. 'Oh, Beth. I'm so sorry.'

'Not your fault.' Her shrug wasn't fooling either of us, but she managed to gather herself. 'So are you still with the same boyfriend from Cornwall?'

It felt safe to return to the table. 'Kit, yup.' The wine in the veins loosened the cue in my hand; my game seemed the better for it.

'It's serious, then?'

'Technically,' I closed one eye to line up my next shot, 'he's not my boyfriend, he's my fiancé. Although I hate that word because it makes me think of some bimbo showing everyone her sparkler.'

'Oh, no, *Laura*,' said Beth, with such disappointment I wondered if she'd misheard me. 'Don't be snobbish or ashamed about love. It's the thing, isn't it? *The* thing in life.' I handed her the cue and raised my eyebrows at her. 'You probably wouldn't think it to look at me because I'm not a typical girly girl, but I've always wanted that, since I was little. It's not weak to want sex and companionship and to be a mum and all those things.' She was right, I realised. I had always thought that the pleasure and comfort I took from my relationship with Kit was somehow . . . not cool. Beth bounced the cue off her palm distractedly. 'I can't ever imagine that happening now.' Her easy confidence of seconds before had vanished. 'He's robbed me of all that. I feel like . . . a porcupine.' She made her fingers into spikes, miming their growth all over her skin. 'How's any man ever going to get past all this?'

'I really hope it'll come.' I sounded feeble at best, patronising at worst. Beth gave a grimace that turned into wide-eyed panic as the bell rang for last orders at the bar.

'Oh, shitting hell,' she said. 'It's never eleven o'clock? I've got about five minutes to get to Liverpool Street.'

'Stay at ours.' My offer was a reflex action, my only concern that Mac had not already commandeered the futon.

The noisy striplit Tube was too crowded for conversation, our carriage still standing room only by the time we got back to Clapham Common. I hated that station at the best of times; rather than two separate platforms, there's one narrow runway straight up the middle, the trains running in deep gullies on either side. There's no wall to lean back against for safety in the rush hour. Its menace only intensifies on the last train and Beth linked arms with me as if for protection as we walked the platform's length.

The flat was dark save for the string of fairylights around the sitting-room bookshelf, Kit's way of letting me know he'd gone to sleep. The bedroom door, as ever, was ajar; the wood had warped over the summers and no longer fitted its frame.

'Pretty,' said Beth, looking out over the balcony at the common below. In the dim light I made up the futon in silence, turning cushions into pillows and the throw into a bedspread.

I used to collect scented candles back then. They were hand-made, the scent was called Blood Roses, they were very expensive, and I never ran out because Kit got me one for birthdays, Christmas, Valentines and anniversaries; so relieved that he didn't have to pass some female test of intuition and intimacy for the perfect gift that he didn't baulk at the price tag. They also neutral-ised the smell from the kebab shop below. I lit one now for Beth. 'Blow it out before bed,' I said. 'It saves you having to feel around for a light switch.' The scent of roses that always lingered in that flat began to swell, like petals were being crushed under our noses. Beth inhaled it.

In our bedroom, Kit breathed out sleep and toothpaste. I fumbled in our shared wardrobe for something for Beth to sleep in and my hands closed around an old T-shirt, which I threw and she caught.

'D'you need a toothbrush?' I whispered. We kept a multipack in the bathroom cabinet for moments like this, ever since I found Mac using my toothbrush to clean white foam off his tongue.

'You're a mind-reader,' she said.

While she brushed her teeth and changed, I glanced into the bag she'd left gaping on the futon. It was almost empty, save for a blue leather purse – she'd bought a new one since the Lizard – a Young Person's Railcard and a creased copy of *Sky* magazine. There was also a clean pair of knickers tucked into a mesh pocket. I switched off the fairylights at the mains, then turned down the corner of the bed and left the Blood Roses candle burning for the girl who carried spare underwear around with her like the rest of us carry our keys.

26

LAURA
19 March 2015

Next door, the builders are chiselling away at something deep in the party wall. *Chip chip chip* go their tools, each beat raising my blood pressure by a degree.

I'm hot with stress. Just like Mac couldn't have one shot of whisky without sinking the entire bottle, my visit to the forbidden corners of the internet triggers a full relapse in me. Knowing it won't do me any good, but powerless against the masochistic compulsion, I type in the URL www.jamiebalcombeisinnocent.co.uk

I hold my breath while the site loads and tell myself that nothing could be worse than what I've already seen this morning, but the home page is just as it has been for the last six months, two days. (Not that I'm counting, as such; I only know because the last time I logged on to Jamie's site was the morning before I found out I was pregnant. After the positive test, I had to protect the babies from the adrenaline spike that viewing the site always created in me.) Gone is the bold assertion, the biography, the contact details and the list of contents. Gone is the years-old flannel about the Criminal Cases Review Board. Gone are the pictures of Jamie and his family, of Jamie on horseback, of Jamie and that award-winning eco-estate he built, of Jamie receiving his Master's Degree in Criminology. The whole thing has been replaced by this message, red letters on a black screen.

JAMIE'S WEBSITE IS BEING UPDATED DUE
TO AN EXCITING NEW DEVELOPMENT
THANK YOU FOR YOUR CONTINUED SUPPORT

I stare at it for a few seconds, then close the window. It can't be that exciting a new development, as it's been up there for half a year now. And anyway, if something *had* happened, the Balcombes' public relations team would have splashed it all over the media. I don't know what's going on. Others might assume that they have admitted defeat, but Jim Balcombe once said that he would fight to the death to clear his son's name and he's still alive.

Who, now, apart from me – or Beth – could overturn Jamie's conviction? It's me he wanted. That's what he was asking for in all those letters, for me to swear an affidavit retracting my statement. The conditions of his probation included a lifelong ban on making contact with his victim but nothing was specified about me.

It can hardly be the case that over a decade and a half later, someone has suddenly remembered something and come forward, so I am inclined to think that this holding page is their version of an old-fashioned test card; something they do to keep the message, the *brand*, alive while they take a break. I can't think what else it might be. The legal machinations of this futile campaign I never grasped in the first place, and the motive goes in and out of clarity depending on my mood.

If it was important, it would come straight to my inbox. The Google alert I set up years ago is for 'Jamie Balcombe + *retrial*'. That search has yielded nothing in fifteen years but its threat never seems to diminish.

The case being retried would force our lives to intersect with Beth's again, and it would wake my sleeping lie. I don't know which is worse.

27

LAURA
19 May 2000

Kit woke up first, and was naked at the bedroom door before I remembered we had company. Thinking more about Beth's potential distress than his modesty, I shot my leg out from under the duvet and made a shepherd's crook of my angled foot, catching him on the shin.

'There's someone in the guest suite,' I whispered.

Kit made a near-miss face – he knew I wouldn't have bothered to warn him if it was Mac – then pulled on his pants and a T-shirt from the floor. 'Who? I didn't hear anyone come in.'

'Don't freak out, but it's Beth. From Cornwall.'

Kit's mouth fell open. 'How – what's *she* doing here? How'd she even find you?'

'She turned up at work last night. I gave her a card at court.' I realised the significance of the words seconds after they were out. Kit's eyes flicked back and forth as though over abacus beads: Beth left the court building after giving evidence. *Ergo* she was not present from the second day onwards. *Ergo* Laura must have spoken to her on the first day. *Ergo* Laura lied to me about where she was, spoke to a witness and jeopardised the case.

'When did you even.'

I caught at his hand. 'We met by accident, in the toilets.' I was whispering but talking fast so he didn't have time to work out that I hadn't used the lavatory while we were there together. Had he worked out that I had sneaked from the hotel room while he slept, he'd have been rightly furious. 'Please don't be cross, it was a sput-of-the-moment thing and I promise you we didn't discuss the case.' Kit gave me a look I recognised from my dad; I'm not angry with you, Laura, I'm just very disappointed. I sat down beside him. 'Look, leaving aside how she found me, I need to tell you why. She came to my work because she was really upset. Jamie Balcombe's been given leave to appeal against his conviction.'

'Whoa.' Kit ran a hand over his unshaven chin. 'If it goes to retrial, you'd better hope nobody saw you talking in the toilets, let alone that you've started having sleepovers with her.' There was contempt under his anger, sending a rush of fear through me. I could tolerate the anger, but I could not bear to lose his esteem. If my talking to Beth made him react like this, he could *never* find out what I said in the witness box.

'You don't have to whisper, I'm awake now,' came Beth's voice from the sitting room. I left Kit half dressed and angry in our bedroom. Beth was on her feet, yawning. When I noticed what she was wearing, I saw that I'd fucked up on yet another level. The T-shirt I'd groped for in the dark was Kit's prized Chile '91 souvenir shirt, threadbare and holey even then but so precious to Kit he barely wore it; and I had let Beth sleep in it.

'Hang on,' I said, throwing the bedcover round her like a cape and drawing it tight around her neck. 'Just stay like this, I'll explain later.'

She obeyed the command without questioning, and I wondered again how complicit we were. When Kit emerged she looked bizarre, her dark curly hair a volcano's plume above a white mountain of bedclothes.

'Hi,' she said shyly to Kit. 'It's nice to see you again. Sorry for monopolising your futon.'

'That's ok,' said Kit mechanically, then disappeared into the bathroom, almost slamming the door behind him.

'He's not really a morning person,' I said, filling the kettle. 'Sorry, d' you mind taking off that T-shirt? It's one of his best, he'll go nuts if he finds out you've slept in it.'

She shrugged off her shroud, looked down at the raggedy garment in puzzlement, then turned away to change into last night's clothes. She had a huge tattoo across her back and shoulders, a pair of vast, spread angel wings, beautiful in their pen-and-ink detail, like something from an eighteenth-century zoology textbook. They flexed with the muscles on her back. I forced myself not to stare. After I'd stashed Chile '91 back into the wardrobe – really, he should have wrapped it in tissue or something if he didn't want people to wear it by accident – I found Beth, arms folded, standing in front of Kit's huge map of the world.

'What are all the lines?' she asked. 'Aeroplane flight paths?'

I'd forgotten what a puzzle it was to the uninitiated. 'It's the path of totality for all the different eclipses of Kit's lifetime,' I explained.

Beth's smile faltered as she lined up her finger next to mine and traced the shadow path across the English Channel and into Europe. 'That's last year,' she said, resting on Cornwall. 'But why all the others?' she asked.

'He follows eclipses around the world; well, we both do now. He's been doing it since he was a kid. We've got trips lined up into the next millennium; the festival movement seems to be growing. Next one's in Zambia in a couple of years, so I should actually get to see it this time.' I heard myself through Beth's ears and could have punched myself in the face. I wanted to pull the map off the wall; it seemed so crass that the worst day of her life was reduced to a souvenir diagram. 'God, I'm sorry, that's so insensitive. Here I am moaning about being clouded out, after what you went through.'

She waved my apology away, but her bottom lip was clamped tight between her teeth. Her attention moved away from the map towards the framed photograph underneath it. Ling had

182

taken it a couple of months before the Lizard. It was the evening of my graduation, and in the photograph Kit and I were entwined on the grass in hired finery; Kit in black tie, me in a pale gold ballgown. Our legs were locked together, our fingers linked, and an empty bottle of champagne lay beside us. We were surrounded by other people but utterly careless of their existence. We will never have another picture taken like it. It's not just the dewy complexions and tight jawlines that can't be recaptured.

'We didn't know anyone was watching,' I said. 'That's why she got such a perfect moment.'

'I want this,' she said, and it was clear she didn't mean the photograph but what it represented. I straightened the photograph on the wall and walked two paces to the kitchen, where I dropped teabags into hot water. 'How long before we know if the appeal's going ahead?' I said.

'Months rather than weeks, apparently.'

'Well,' I said. 'You know where I am now.'

'Indeed I do,' she said, looking around her, like she was trying to memorise everything about our little flat.

Kit darted from the bathroom to the bedroom, emerging seconds later in his work clothes: Adidas Gazelles, jeans and a lumberjack shirt; the young man's equivalent of corduroy and patched elbows. He grabbed a slice of dry bread from the worktop and shoved it between his teeth.

Beth picked up another photograph, one of a rainbow over the common, a seven-lane highway in the sky.

'Where did you buy this?' she asked me.

'Kit took it,' I said.

'Seriously?' said Beth. 'What with? I know a bit about photography, I did it for art foundation.'

'An old Nikon Prime,' he said, thawing at last. 'They've fallen out of favour, but I still love them.'

'The Prime's a good machine,' she agreed. 'Have you got a super-telephoto lens? They're really good for shooting the sky.'

'Yeah, well, one day, when we win the lottery,' he said. He wasn't friendly, exactly, but at least he wasn't rude. 'Laura, come on, we're going to be late.'

I was in and out of the shower in ninety seconds. I Febrezed a dress with no visible stains and then gave my hair the same treatment. Kit was halfway down the first flight, his tuts amplified by the echo chamber of the narrow stairwell.

'Let's go,' I said to Beth, stepping into my work shoes.

'I don't suppose I could grab a shower?' she asked. I looked at the clock. Ten to nine. I was cutting it fine as it was.

'It's ok, I can see myself out.'

My hesitation was momentary. I wouldn't normally leave a virtual stranger alone in my flat; but then I reasoned that this friendship was on fast-forward.

'Sure,' I said. 'There's a spare towel on the back of the bathroom door. Hang it over the banister when you're done.'

I caught up with Kit at the entrance to Clapham Common Tube.

'Where's Beth?' he said, looking over my shoulder.

'Having a shower.'

He raised his eyebrows in reply.

When I got in at half past five, Beth had done the washing up and tidied the flat so thoroughly that it almost looked rearranged, although a glance at the bookshelves told me they were ordered as they had been that morning, only straight and somehow cleaner. The bookshelves disturbed me even more than the gleaming glasses or the made bed; I got the feeling she'd been through them too, read them, like she was trying to read *us*. At six, a text came through.

Hope you don't mind me doing the full Cinderella on your flat. My way of saying thank you, for all of it.

You don't need to thank us – but thank you, I sent back.

When Kit got home, very late, with a sagging satchel of essays to be marked, he interpreted the tidying as a peace offering on my part for bringing Beth into our flat, and I didn't correct him.

28

LAURA
20 May 2000

I would have been easy to track down in those days. Langrishe is an unusual name; I have never chanced across another Langrishe. When the letter came, a jaunty diagonal on the tatty doormat, I knew what it was at once; not just its origin, although the prison frank told me he was now in Wormwood Scrubs, but its contents. He could have only one reason to write. He had written on – oh, irony – lined, yellow legal paper.

Dear Laura,
I write this in my cell at Wormwood Scrubs. Next door there's a serial child rapist. Last week he threatened a female warden's life with a razorblade embedded into an old toothbrush handle. This is my life now. These are the kind of men you have condemned me to live with. The only thing keeping me going, apart from Antonia and my family on the outside, is that I know I do not deserve to be here, and that I will doubtless be released when my name is cleared.

Why, Laura? I am still at a loss as to why you lied in the witness box at my trial. You know that you did not hear my accuser say no. I know you know it. You might have fooled the jury, you might even have persuaded my accuser you were telling the truth; but you and I know. How do you live with yourself?

You will have heard by now that we are appealing the verdict. I am confident that before long we will meet again in a court of law, and this time my counsel will expose you. Isn't it better to do the decent thing now, to contact the police, or any of my repre-sentatives, and correct your testimony rather than have it happen in court? Of course there will be repercussions for you. But wher-ever they send you cannot be as bleak as where I find myself now. I will keep writing to you. If I write to you enough – and I have time on my hands – I believe the gravity of what you have done will sink in. I saw you in Cornwall and I saw you in court. I recog-nise passion and principle when I see it, and I am sorry that these qualities have guided you to the wrong conclusion. But your conscience must be pricking and I make no apology for exploiting that. So please, please: take back your lies and give me back my freedom.

Yours,

Jamie Balcombe

I thudded on to the stairs so hard my hipbones hurt. The arro-gance of the man hit me first; that he had the audacity to talk in terms of things we both knew, when I had *been* there, when I had *seen* it. In the confident tone of his letter I sensed again that knife-edge charm. I thought then that he must be in breach of at least one law, to write to a witness. I looked into it later that week, making furtive, expensive phone-box calls to witness care and the probation service, and found that this kind of thing happens a lot more often than you'd think. Writing to a witness is only a crime if there is intimidation, and he was too clever to threaten me outright. He must have known that I was too scared of exposure to take it to the authorities, anyway. Checking outgoing post, I learned, is a random process, and they're looking for breaches of security – drugs, or escape – rather than talking about the case. I suppose if they censored every convict who protested his inno-cence, the letters would dry up pretty quickly. Maybe if I had kept all the letters, their frequency and volume might have added up to some kind of harassment or intimidation, but all I wanted then was to get rid of them before Kit could see them.

Back then, I could easily differentiate between Jamie being right, which he technically was, and his righteousness, which I didn't believe he was entitled to.

I opened the street door and left it on the latch. Tiptoeing in bare feet across the filthy pavements of Clapham Common Southside, I put the letter in a public wastebin, wedging it between an empty Starbucks cup and a newspaper. All weekend, I felt its presence in the street outside. I didn't relax until the Tuesday morning, when the refuse lorry came at dawn. I watched from my balcony as the men in overalls tipped one bin after another into the dustcart. I convinced myself that I saw the yellow paper, bright as a lie, churning over and over in the rubbish as the maw of the refuse lorry chomped down on the indigestible truth.

29

KIT
19 March 2015

After the viral video debacle, I need a gesture, something to appease Laura. The equivalent of coming home with champagne and flowers, the romantic flair I have so often been told I'm sorely lacking. Darren's film captured my two distinguishing features. The first, my Chile '91 T-shirt, is now rolled tightly into a side pocket of my rucksack, where it will stay until I go back to London. The second, the ginger beard, makes me instantly recognisable even in this Viking country. Getting rid of it is the best way I can think of to regain my anonymity. That's how I find myself in Me Time, the *Princess Celeste*'s on-board beauty parlour. It's a terrifying, unportholed room that smells of female hair and alien chemicals. There's a single sink with a shower attachment and one of those dips for your neck that looks like an executioner's block. An old lady with diamonds on her fingers and orthopaedic shoes on her feet flicks through *Hello!* magazine under an old-fashioned space-helmet hairdryer.

There's a price list on the counter and I run my eyes over her list of 'services'. I don't need a cut and finish and I don't need to know what a Hollywood wax is to be sure I don't want one. I can't see what I want. I hate going off-menu in any context, but desperate needs call for desperate measures.

'Can I help you?' A woman about my mum's age comes bust-first through a louvre door. Her hair is short, swishy and streaked with plums and burgundies.

'I don't suppose you do an old-fashioned wet shave?'

'Not the old-fashioned kind.' She smiles kindly. 'Cut-throat razors and choppy seas aren't the best combination. If you want to take the fuzz off I can sort you out with the clippers and a safety razor, though.'

'How much?'

She looks me up and down, clearly sensing my desperation.

'Thirty pounds.' Her smile now seems edged with mockery, and instead of the *Thirty fucking quid* that backs up behind my teeth, I say, 'Yes please.'

She tucks me into a huge bib and ties a towel under my neck; it would almost be comforting in other circumstances. At the clippers' touch, coarse red hair floats to the floor. I'm expecting a can of shaving foam but she lathers me up with a shaving brush and soap, so exactly like the one my dad used to use that suddenly I'm back in Chile for the '91 eclipse. Mac and I were about twelve, with perhaps three whiskers between us but determined to start shaving, and when Dad was passed out on the beach, we went through his sponge bag and helped ourselves to his things. We were half blind with laughter as we lathered up, then used his crappy blunt old razor, still thick with his greying bristles, and cut our baby faces to ribbons.

This memory whirls me along to another; we had a lot of firsts on our travels with Dad, usually while he was out drinking or sleeping off the drink in a hotel room or a beach cabin or a trailer somewhere. The following summer we smoked our first cigarette, stolen from the soft paper packs of American Spirit he used to smoke; the tobacco was organic, which Dad took to mean it was virtually a vitamin pill. We weren't laughing when we did this; it was deadly serious, a fumbled ritual that then became funny when we realised that if you wanted to light the cigarette you had to actually breathe in at the same time. This accomplished, I took the first drag and nearly fainted. Mac said it went down better than oxygen.

The year after that was Brazil '94. We were fourteen. Dad made a road trip of it; we drove from the airport in New Mexico all the way down to Brazil, where his old friends were staying. En route we got drunk for the first time, stealing a litre bottle of Whyte and Mackay rum I thought would last us the week. It was gone in half an hour. We both threw up, and only Mac went back for more. I have never touched it since.

That trip I ought to have had my first kiss, too, but Mac had other plans.

The rum was gone, but there was a bottle of gin, which we took down to the beach where all the eclipse chasers' kids lit bonfires and drank at night. They were mostly Americans and our London accents were an aphrodisiac. While Mac told a sexy, punky seventeen-year-old *I can tell you're a very spiritual person*, I got quietly chatting to a girl called Ashley who had the kind of slow-burn prettiness it was easy to dismiss at first glance. She was sharp, funny, and when she asked me if I wanted to 'make out' I said yes, in a minute, and then immediately disappeared to take a leak at the bottom of the cliff. When I came back to the fire, Ashley was on her back in the dunes, my twin writhing on top of her with his hand in her bra. Even then, his sex life was already a series of overlapping conjugations. His justification for his serial infidelities was that time is a great absolver; the further in the past the deed, the easier it is to live with until it gets to the point where you barely even did it. Knowing that guilt will one day fade, he says, makes it fade all the faster.

He's always been full of shit.

Later, after Ashley had shaken off the sand and gone home to her parents, Mac didn't understand why I was so angry; he actually said he was 'breaking Ashley in' for me. As though any girl would want me after she'd been with him. It's a sour memory that makes me wince, and the hairdresser nicks my skin. My eyes are flung open; in the mirror, a red rivulet runs between white crevasses.

'Ooh, silly boy,' says the hairdresser. I hold still while she scrapes what's left of my beard off my face. Christopher Smith is gone and Kit McCall is revealed. Beth will be expecting a bearded man and I hope that if we do come face to face, this will buy me a few seconds to read her cues.

The cut on my cheek bleeds whenever I smile and when I send Laura my apology in the form of a selfie, I show her my good side.

Back on the deck, the sun is setting over Tórshavn. I decide to stay here rather than take my chances in the bars below. The ship is shiny; reflection is everywhere, in glass doors, polished brass and curved chrome. Every time I catch myself in one of these ersatz mirrors I'm in the same position, stroking my chin, a parody of a philosopher, or the professor I never became.

30

LAURA
28 May 2000

It was a Friday evening, the first real summer night of the year. On the pavement below, the pubs and cafés set out tables and chairs, as though an afternoon's sunshine was enough to turn Clapham Common Southside into the Champs-Élysées. Pavement smokers took in smog along with their nicotine, but a clean zephyr carried fresh air over the treetops into our stuffy flat. I was on the balcony, trying to give Kit some space; the twins had had a blazing row the previous day when Kit had refused to lend Mac money. Kit was on the futon, moping over his laptop, when the house phone rang. I picked up.

'It's all kicked off,' said Beth by way of hello. 'I need to see you.'

'Sure,' I said, untangling my foot from the mess of wires that pooled around the base of the telephone table; if you weren't careful, one mistimed tread would pull the plug on the phone, the internet and the TV all at once. 'I don't think we're doing anything at the weekend.'

Her voice shrank. 'I'm just at the Tube.'

'Clapham Common Tube?' The sharpness in my voice made Kit look up from his screen.

'Yes,' she said. 'Sorry. I had to get away.'

'You'd better come up, then.'

Kit sighed and stretched so that the laptop tilted. 'Mac?'

'No,' I said. 'Um, Beth Taylor's outside.'

Resignation turned to concern. 'Something to do with the appeal? Why's she have to come here?'

'I don't know,' I said, convinced my time was up. I had read up about the appeal process. It seemed that lodging an appeal was just what you did, a given for any guilty party of significant means. The chances of the case being retried were tiny. The prospect of repeating the ordeal had become a cloud in the sky rather than an event on the horizon. Now she was back, which could only mean it was really happening. I felt sick as I buzzed Beth in. The confessional urge writhed inside me again but desperate now and dark; the imperative no longer one of unburdening but of damage limitation. Tell him you lied in the witness box, I thought, as her footsteps sounded on the stairs. If he's going to find out, better that it comes from you. Tell him, before she gets here. But I couldn't get the words out.

Beth was flushed by the time she got to us; sweat had pulled her loose curls tight into corkscrews around her face.

'What's happened?' I said. 'Is the appeal going ahead?'

'No,' she panted. 'I mean, I don't know. It's too soon to know about the formal appeal. But they're waging fucking war on me.' There was a glass of half-finished wine on the worktop and she looked at it with undisguised longing.

'Have it,' I said. She downed it in one, then looked at Kit's laptop. 'Is that online right now?' Kit nodded, looking as mystified, if not as nervous, as I felt. 'Can you type jamiebalcombeisinnocent.co.uk?'

Kit minimised the half-dozen screens he'd been flicking between and called up the right page. The Balcombes had clearly been busy.

JAMIE BALCOMBE WAS WRONGLY CONVICTED OF RAPE ON 20TH APRIL 2000. THIS WEBSITE IS RUN BY HIS FAMILY AND FRIENDS, WHO WILL CONTINUE TO FIGHT TO CLEAR HIS NAME.

'Can they say that?' I asked. 'They're basically saying you lied. Isn't that libellous?'

'No,' said Beth tautly. 'You can't libel someone anonymous. They can say what they like about me.'

On 20th MAY 2000, Mr Donald Imrie of Imrie and Cunningham Chambers, Bedford Row, was granted leave to appeal the verdict. They are confident that the verdict will soon be overturned.

'Keep going, it gets better,' said Beth, reading over our shoulders. 'Can I have some more wine?'

'Go for it.'

From the outset we would like to state that this website does not seek to undermine the seriousness of rape or trivialise the suffering that rape victims suffer. Had Jamie committed the crime, we believe that a five-year sentence would not be sufficient. But we are stating that he did not commit the crime of rape at all. Furthermore, we recognise and acknowledge that in rape cases the anonymity of the victim is a fundamental legal principle that should be upheld and respected. However, we would like to see that right extended to men, too. This is something to which Jamie intends to devote his energy when his name is cleared.

We are unable to comment on whether we will counter-sue Jamie's accuser in the likely event of the verdict being overturned. We have always maintained that her history of mental instability makes her vulnerable and it is more important that Jamie's name is cleared than that an already troubled young woman is put through the criminal justice system when her problem is a medical one. We are willing to help her obtain the professional counselling she needs to help her come to terms with what she has done, and to address the deep-rooted problems that led her to make this false accusation in the first place.

'Bollocks, more like,' said Beth. 'If they get him off, they'll bring charges against me before the handcuffs are even off him.'

I didn't doubt it, but had to concede that the false note of concern was a public relations masterstroke.

'You finished reading?' I asked Kit, who was staring slackly at the screen. He nodded.

There was a series of tabs down one side: *Offer Jamie Your Support, Appeal for Information, Gallery, Jamie's Career. Media Enquiries* told the reader that in addition to instructing a new legal team, the family had hired a publicist. 'No stone unturned,' I said. I clicked on *Antonia's Letter*. The accompanying photograph showed her and Jamie at a wedding; she had a few flakes of confetti in her hair.

> *Thank you for visiting Jamie's website. My name is Antonia Tranter and I am Jamie's fiancée. We have been together for two years. The past year has been hard for me. It is not easy to hear that your fiancé has been unfaithful; but that is nothing compared to the nightmare that followed, when this mistake has had him painted as a monster. Like all who know and love Jamie, I am convinced of his innocence.*
>
> *On these pages we have set out, as clearly as we can, reasons why this has been a gross miscarriage of justice. We'd be grateful if you could share our message as far and wide as possible.*
>
> *In particular, we are looking for witnesses who may have seen Jamie and the young lady in question prior to the event. However insignificant you feel it may be, or if your conscience is troubling you, please write to me via the website and help us get justice for Jamie.*

'He's got her brainwashed,' I said.

Beth chewed on a cuticle. 'Have you found my starring role yet?'

Kit and I looked up together.

'Click on the tab that says, "Judge for yourself",' she said.

Mr Justice Frenchay praised Jamie's accuser in court for her strength of character in coming forward, and in his summing-up painted her as a shrinking violet. The judge also glossed over her history of mental instability. Whilst we sympathise with anyone who suffers from mental illness, we feel that the jury did not take into consideration that this makes her an unreliable witness.

I couldn't resist looking at Beth.

'D'you want to know about my history of mental illness?' she said. 'When I was sixteen my grandparents were killed in an accident. I was heartbroken.' She looked at me, appealing to my own loss, and my heart went out to her. 'I couldn't sleep, so my GP gave me a couple of weeks' Valium. That's literally it. Just good old-fashioned grief. In the witness box, they made it sound like I'd been in and out of asylums ever since.'

These photographs show Jamie's accuser out 'partying'. We have of course disguised her identity in accordance with the law. Is this the sober, conservative girl who appeared in the witness box? Or is this a free spirit, a hedonistic, party-loving girl who is open to all experiences, for whom casual sex at a music festival is all part of a weekend's fun? We believe that, had the jury been allowed to see these pictures, the verdict would have been very different. Who do you believe? The diligent, studious young man with no record of violence, or the girl with a history of mental illness, who allows herself to be photographed like this?

In the pictures, Beth's face was pixelated but the hair was clearly identifiable. One showed her licking a male friend's face in a nightclub; not the most elegant photograph, but everyone our age had done worse. The other had blacked out her whole head but showed her whole body; she looked great in a black bustier, hotpants and cowboy boots, with a bottle of tequila between her breasts. She was photographed side-on, with one shoulder turned towards the camera. The angel wing tattoo had not been disguised.

'This is *bullshit*,' I said. 'Anyone who knew you could recognise you from that.'

'You don't need to tell me,' said Beth bitterly. 'They might as well have taken out a front-page advert in *The Times*. You should've been there when I went shopping with Mum yesterday. It was like the parting of the Red Sea, just backs and shoulders in every aisle.'

'Who gave them the pictures?' asked Kit.

'One of my oldest friends, can you believe it? The first one's in a club in Nottingham, the second one was a summer job selling tequila shots at a local festival. I remember Tess taking the pictures. She's the last person I'd have thought would do something like this. She's broken my heart.'

She started to cry and something inside me lashed out at Tess.

'Oh, Beth, that's shitty.' I remembered a picture Ling had once taken of me; raving in jeans and a bikini top, mouth open, pill on my tongue, a gift to anyone who wanted to smear me. We'd had it stuck on the fridge in our old flat, taken down when parents came to visit. When you're young, you don't think about consequences.

'You said they can't libel an anonymous person but you're not anonymous in these, are you?' said Kit. He was speaking almost to himself, chewing over the legal ramifications. 'They must be in contempt of court, identifying you like this. This is a massive own goal, surely. Have you got legal representation?'

'Yes,' sighed Beth. 'My dad's geriatric solicitor is on the case. He reckons he can get them to take it down, but the damage is done. People know.' She flopped down suddenly. 'The ripples just keep coming and coming at me. I don't just mean the appeal hanging over my head although that's bad enough, but the things people say. Yesterday, in the Co-op, a girl I've known since I was *four* said that if I went on my own to a festival what did I expect? And this woman's our age, she's normal, she drinks, she's no virgin. She's the last person you'd think.' She looked around our flat, at the eclipse map, at me, at Kit, and then fixed her eyes on the view across the common. 'Right now, this flat feels like the only place in the world where I can be believed. I need to get the fuck away from there. I can't bear to watch my parents staying strong for me when they're falling apart over all this. And I can't make a life there, not now.'

'Stay here,' I said. 'For a couple of nights. Till it dies down.'

'Do you mean it?' she said. You'd have to know Kit as well as I did to tell what he was thinking: the closer you get to this girl, the more dangerous it is at a retrial. But quickly, before Beth could see, he flicked on a grin and I knew he wouldn't rescind the invitation. He was doing it for me, not her.

'Sure,' Kit said, closing the laptop with a bang.

'Oh, *thank* you,' said Beth. Her tears stopped abruptly, like her stopcock had been turned off. 'What would I do without you both?'

31

LAURA
30 May 2000

'What are you doing?' whispered Beth, as I tiptoed up the stairs and tried to sneak past her. It was a Saturday, and we'd all slept in. I thought I'd heard the clatter of the letterbox from my pillow. Since Jamie had starting writing to me, I'd got into the habit of being first down to the doormat in the mornings. I couldn't risk Kit intercepting a letter from prison. I sat on the edge of the futon, a bank statement and a pizza leaflet in my hand.

'I just like to be the first one to pick the letters up,' I said. 'It's just a thing I do. Like a routine. A sort of comfort thing.' I was over-explaining it, while not explaining it at all.

'Oh,' said Beth. 'That's . . . nice.'

'Morning.' Kit hovered awkwardly in our bedroom doorway. I'd tried to shut it behind me but it had bounced on its hinge as usual. 'God, it's late.'

'I'm going into London today,' said Beth purposefully. 'Things to do.'

'I thought you didn't know anyone in the city,' said Kit.

'I don't.' She smiled. 'That's the appeal. I can do normal things without worrying that everyone's talking about me. I can walk up Oxford Street and not know a soul.'

After breakfast, we leaned over the balcony and watched her

go, in a daffodil sundress and those same old silver Chipies she'd worn in Cornwall,

'How long does she expect to stay here?' said Kit, grinning through gritted teeth as she turned and waved.

'I have no idea,' I said. 'But you can see how much she needs us. And I can't chuck her out after what she's been through.' Kit knuckled his eyes. 'I'm not asking you to. It's just, the timing's not ideal, is it, to have to stand on ceremony in my own house.'

'She's only been here like four nights. God, Kit. You act like I'm choosing her over you. It's not an either/or situation.'

Anger rippled beneath the surface of his face. 'Isn't it?'

I exploded. 'D'you know what? If you don't develop a bit of fucking empathy, it might be.'

The plum half-moons under his bloodshot eyes seemed to darken and I wished I could take it back. I took in for the first time how awful he was looking, how uncared-for. He'd had the same T-shirt on for days and his hair was starting to curl at the collar. He turned his gaze towards the common.

I was about to apologise when he blurted, 'If we were identical, do you think I'd be like Mac? Like Dad was?'

The apparent non-sequitur showed me that our thoughts were travelling in parallel and I forced myself to ride his tracks. The moment to talk about this me-or-her nonsense with Beth was gone, never to return. 'God, Kit, I don't know. I don't know enough about it.'

'I almost wish I was, sometimes. No, I don't mean that. I wish I could get inside him and understand how his brain works. If he goes the same way as Dad . . .'

'Hey.' I took Kit's hand. His palm was warm, dry and smooth. 'Mac's *years* younger than your dad was, and he's got us. We'll catch this in time.' We stood in silence for a while, Kit staring into the middle distance, me looking down over the balcony. Red buses owned the street.

'About Beth,' I said tentatively, when I thought enough time had passed. He didn't say anything but I felt his irritation in the squeeze of his hand on mine. 'No, hear me out. I wanted to say, we'll just indulge her for the day, ok? Come on. You've got your lost cause. I've got mine.'

I'd meant it as a joke, but I'd offended him.

'How can you even compare the two? Mac's my *twin*.'

I knew then that there was no way Kit would understand the bond I felt with Beth. You can't compete with blood.

'Hi honey, I'm home!' Beth called, laughing her way up the stairs. Kit confined his irritation to an eye-roll. She was lightly caked in the grime of the city and laden with 7–11 carrier bags full of onions, tins and wine. A stick of fresh lemongrass poked from the top.

'I'm cooking for you,' she said. 'You haven't lived until you've tried my Thai green curry.'

'I love Thai,' I said, loud enough to cover Kit's snapped, 'There's really no need.'

If Beth heard him, she chose to ignore it. 'We've got something to celebrate. My solicitor called. Those pictures are down off the site.'

'That's great news,' I said.

'There's a chance we can sue, but I don't know if I want that. I'd rather put it behind me, you know? I've had enough of lawyers and courts.' She laid out her ingredients – jasmine rice, coconut milk, a knuckle of root ginger, three fat chicken breasts – on our worktop, then rummaged in her vast handbag. 'I got you something,' she said, suddenly shy. 'To say thank you for letting me stay, and –,' she looked meaningfully at me, – 'generally going the extra mile. You first.' She produced a little gift-wrapped box the size of a house brick and watched expectantly as I opened it. I could tell before the paper was off that it was a Blood Roses candle. In 1999 scented candles were not quite the ubiquitous gift items that they are today, and you couldn't get the Blood Roses range anywhere but an obscure little shop in Marylebone. Beth must have memorised the label and then done her homework. She had bought me a gift set: three fat candles, their wicks untrimmed. The sweet, heady scent filled the room before they were even lit. They must have cost her a hundred pounds.

'Wow,' I said. 'Thank you.'

'Well, it's my fault, really. I fell asleep with it on, that first night, burned it right down to the wick.' She nodded apologetically at the empty glass on the mantel. 'I love flames, and it's such a sooth-ing smell.'

Kit narrowed his eyes: it was his job to keep in me in Blood Roses candles. I returned his sulky look; Beth wasn't to know she had trodden on his toes.

'And Kit, this is for you. It's only second-hand, but it's in perfect condition.' It was the telephoto lens she'd been talking about. 'They're *amazing* in low light.' She'd shocked his manners out of him. 'You've got one already.' Her voice seemed to hover over the words.

'No, no I haven't.' His voice was flat. 'Thank you.'

'I think Kit just feels awkward because . . .' I looked at him, but he was impossible to read. 'I think Kit just feels awkward because you don't need to keep thanking us.'

'Yes,' said Kit. 'All we did in court was say what we saw.' His tone was detached, which meant he was boiling with embarrassment.

'No,' said Beth. 'You saved me. In more ways than one. You *saved* me.' An awkward silence trailed in the wake of her words, broken only when Beth shook herself like a dog flinging off water. 'Well!' she said brightly. 'Supper won't cook itself.'

While she chopped and stirred, I put some music on and lined up the trio of candles on the mantel. After a couple of glasses of wine, Kit took out the lens and screwed it on to his camera, unable to hide his delight at his new toy.

We lay face-to-face in our bedroom that never truly got dark, hoping our whispers wouldn't travel through the gap in the door.

'How much was that lens?' I said. 'Like, a hundred quid?'

'More like a grand. They go for up to three, brand new.'

'*What?*'

'I know,' he said. 'It's like she's trying to *buy* our support.'

I tried to look in his eyes but they were only glitter in the half-light. 'How can she buy us? We've already testified.'

'She's trying to keep us onside, then. I'm worried something's going to come up in this appeal and it'll go to a retrial. And that by having her here, we'll invalidate the whole thing.'

'Only if we tell people she's been here. Unless they've got like a private detective on her case, who's to know?'

Kit tensed and I could tell it was an effort for him not to raise his voice above a whisper. 'This is what I was afraid of. Getting drawn into a whole mess of lies. One lie always needs another one, and then another one. It's started already. Unless you're totally straight from the beginning, you're fucked.' I froze and then remembered he wasn't talking about my secret trip to the courtroom. I put a hand on Kit's chest to calm him down.

'She just needs a friend.'

'Laura.' He caught hold of my hand. 'How can you seriously expect to forge a genuine friendship with someone you met like that? It'll always be hanging over you. You've already done more than your bit.'

'What do you want?' My voice threatened to break its whisper. 'You want me to tell her to go?'

Kit shushed me, then said, 'Honestly? Yes. I haven't got room in my head for this, let alone my flat. It's more than I can do to look after my brother. I'm not like you. I don't want to rescue every waif and stray.'

'You said,' it was an effort to control my voice, '*you* said, when we were first together, that that was what you loved about me. That I cared about things. And issues. And people.'

He flopped back on the pillow. 'Yes,' he said. 'I did. I *do*. But sometimes I wish you gave the same attention to things a bit closer to home as you to do your *causes*.'

He kept his back to me. In the room next door, Beth rolled over, and the futon creaked a reminder, as if any were needed, that we were not alone.

32

KIT
20 March 2015

The sky through the porthole is a wash of light grey nothing, and pre-emptive disappointment takes away my appetite. I didn't go looking for Beth after all and this morning I'm tortured by the missed opportunity.

My bag is heavy with my camera kit; lens filters, rain covers, my sturdiest tripod. For a moment I'm tempted to leave the whole thing in my cabin, sod the lot. I'm weirdly close to tears. Everything always feels more highly charged in the hours before an eclipse.

They open the *Princess Celeste*'s restaurant just before dawn, serving a continental breakfast. I eat breakfast, or rather drink coffee on an empty stomach, with Jeff Drake. He doesn't comment on my new clean-shaven look. Why would he? He knew me looking this way for years and has only seen me bearded the once. And if he's seen my performance on video, he doesn't comment on it. He has the distracted air of a man whose mind is on higher things. It is his task to decide where our best chances are.

'Where do you reckon?' I say, taking a slurp of coffee. I must find out the blend for Mac. 'North side of the island? South?'

'I'm afraid you need to rephrase the question, Christopher,' he says, sprinkling sugar over a halved grapefruit. 'It's more a case of the least-worst scenario. Predicting the movements of heavenly bodies is a piece of cake compared to forecasting weather.'

There's a certain look that eclipse chasers get on their faces when it's going to be clouded out; an enforced jollity, a determination to enjoy whatever happens. But you can't mask disappointment like that, no matter how hard you try. I feel a kind of responsibility for Richard; it's my fault he's here. I feel like it's the least I could do, to bust the clouds for him.

'Sometimes the clouds part at just the right moment,' I say as we trudge down the gangplank to the waiting buses. 'You only need the smallest gap in the cloud for it to all come good.' I'm talking for my benefit as much as his.

The harbourside is thronged with people, a press of bodies tighter than at any music festival. Above the sea of heads, and across a gridlocked road, I can see six yellow coaches with *Princess Celeste* cards in the windows. I steer Richard towards the coaches; the location has been decided at the last minute and cards in the window tell us we're bound for halfway up Húsareyn, a mountain overlooking Tórshavn harbour. Only amateurs go for the top of a mountain. Too high and the cloud comes down to meet you, obscuring the view from the ground.

With so much human activity it's weird how one other person's eyes, their stare, can make itself felt, but there's something, a kind of heat, at my left shoulder, or perhaps I just see it out of the corner of my eye. I turn my head slowly with a familiar sense of foreboding, but it's just a middle-aged man with a shiny head and flip-down shades clipped on to thick glasses. He breaks into a broad grin.

'It *is* you,' he says. I scan through all the people I've met on the boat and I'm pretty sure that he isn't from the *Celeste*. 'Oh, look at you, panicking about remembering my name,' he says. His American, midwest sort of accent could place him almost anywhere in the States and for a second I wonder if he's someone I knew as a teenager. Someone's dad, come to collect a drunken teenager from an illicit campfire.

'I'm sorry, you'll have to remind me . . .' I begin.

'Don't worry, we don't know each other,' says the man.

'Haha,' says Richard. 'Your fame goes before you. Another one who's seen your video. That's weird, though, that you recognised him *without* the beard.'

The man looks nonplussed. 'What video? No, there was a woman walking around with a photograph of you, looking for you. Taken a few years back, but I'm sure it was you.'

My heart flips like a landed fish. She isn't looking for Christopher. She's looking for Kit. She's here, and she's going to fuck up my eclipse. 'I don't think so,' I say. 'It must be someone else.'

He studies me. 'Either that or you've got a twin brother some-where.' I flinch at that, but I can tell from his face that it's just a turn of phrase.

'What did she look like?' I wonder if Richard can detect the wobble in my voice.

'About your age, I'd say. Pretty. Dark hair. She went off on one of those pink coaches.'

He nods towards the harbour. I see the Calpol-coloured paint-work shining, just proud of the harbour's curve.

Fuck fuck fuck fuck *fuck*.

'Well, I'll keep my eyes open,' I say. 'Have a good eclipse, yeah?'

'Good luck!' says the man, his face shining with idiot hopeful-ness, and vanishes into the crowd.

If I'm right, and that was Beth with the photograph, she's liter-ally boarding right now. Something inside snaps, almost audibly.

'Be right back,' I say to Richard. 'Hold this.' I give him my bag, but keep my camera. It dangles in my hand like a slingshot.

'What do you mean, be right back? It's not a moveable feast – Chris! *Chris!*' Richard's voice trails away as I press through the tight pack of bodies. Every time I push past someone there's the zippy nylon swish of contact with another person in sensible waterproof clothing. Is this what it's like for Laura? Knowing you're behaving like an idiot, powerless to stop?

I try the nearest pink coach first. 'Are you with this party?' The woman with the clipboard has bubble-permed grey hair and a smile that's tipping into alarm.

'No, but I need to check if someone's on the coach.'

'This is strictly pre-booked tickets only.' She looks over her shoulder to the coach driver for back-up but he's about ninety. 'Sir, I'm going to have to ask you to—' She puts up zero resistance as I barge past her.

'I only need to see if she's there.' I stand in the gangway of the coach. One sweep of the passengers tells me Beth is not among them. 'Sorry. Thank you,' I say, bowling back down the stairs, and the murmurs start up: 'Is he on something?' 'Was that a British accent?' 'Should we call someone about him?'

The second pink coach, parked directly behind, is still board-ing. I bypass the front entrance for a second door halfway down and this time I don't even pretend to ask. White heads bob in concern. There's no one here under sixty-five.

Someone's radioed or phoned ahead because they're wise to me on the third coach. Two men, arms folded, wait for me at the entrance. I'm forced to shout up the stairs.

'Beth!' My voice is loud enough to travel the length of the coach; it's loud enough to cause avalanches. '*Beth!* I've had enough! Come on! You win! Let's have it out! I've had enough!' The door swings shut in my face. 'I'm losing my mind out here,' I shout through the glass. I press my face against the door and wait. If she's on board, she must have heard me; and if she's here, she's here for me. She will come. 'I'm losing my fucking mind,' I repeat in a whisper.

The driver revs the engine and my face judders against the body of the coach, shocking me back to my senses. I sprint the length of the emptying harbour, and find Richard, on the yellow coach, my bag on the seat next to him.

'What the *bloody hell* was all that about?'

'We need to switch seats,' I say, hoisting my bag and lugging it to the back row. From here, I can see every passenger board. My muscles are locked in rigor. I am braced for confrontation. I *hate* thinking like this, but if anything happens to Beth today, she has brought it on herself by following me here.

No, that's not fair. Whatever her faults, Beth went through hell too. It's Jamie Balcombe I should be angry with; if, indeed, the buck stops there.

33

LAURA

25 June 2000

'Are you sure you haven't moved it?' said Kit, for the twentieth time that hour. 'It's a little canister of film in a plastic tube. About so big. It can't have just disappeared. It was literally *here*.' He pointed to the worktop.

'It's probably gone down the back of the cooker or something,' I said. 'Can you not wait till we're finished?' I was sitting on a kitchen chair with a towel around my neck and my hair parted in the centre, while Beth, in latex gloves, painted a violet paste of hydrogen peroxide on to my roots. I felt the familiar caustic tingle as the bleach lifted the colour from my hair. I'd already had a go at hers, turning a couple of dark curls bright purple at the ends.

The radio was on so loud that the water in our glasses jumped with the beat, but her phone, propped on a bookshelf, was still audible. The screen glowed green, the word HOME spelled out in dot matrix. She gave it a look, pinched her face slightly, and ignored it. Kit raised his eyebrows, but said nothing.

The louder Beth and I sang along, the more Kit winced at our banality. If he was going to act like a moody dad, then I would take a childish pleasure in annoying him.

'Ignore him,' I said. 'He only likes music that's about *ideas*.'

Beth gave it an 'I'm not getting involved' look and scooped another measure of bleach into her mixing bowl.

Kit's frown deepened as he returned to his main project in the living room. As if the tragedy of the missing canister wasn't bad enough, now something had gone wrong with the fan on his laptop. He shone his lamp on the exposed motherboard and picked up a tiny screwdriver. When Beth's phone rang out again, his shoulders raised another inch.

'Do you maybe need to get that?' I said, after it started up for the fourth time.

Beth splayed her fingers in hesitation for a couple of rings before saying, 'God, ok, give me a *chance*,' but she was snapping at the phone, not me. She rolled the gloves off her fingers, threw them in the kitchen sink and picked up the phone. She closed the flat door behind her before accepting the call with a defensive, 'Hello?' Her footsteps descended the stairs and she was out of earshot.

'What's all that about?' I said to Kit. 'Do you think she's gone off without telling them where she is or something?' I could see why she'd moved back home, but still to be answerable to one's parents aged twenty-one seemed to me a further indignity.

'Could be,' said Kit, blowing dust off a piece of circuitry. 'I was thinking more along the lines of something bad might have happened to one of them.'

'Oh, God, you're right,' I said. 'As if she hasn't got enough on her—'

We were interrupted by Beth's voice, sharp and shrill through the closed door.

'She deserves it! She's actually lucky she wasn't *in* the car when I did it!'

Kit raised his eyebrows. I crept across the carpet, put my ear to the front door and strained to listen.

'No, I didn't mean it, of course I wouldn't. You know I wouldn't have.'

I put my hand flat against the door and felt my pulse in my fingertips.

'Well, I'm sorry for embarrassing you,' said Beth eventually. 'I wasn't thinking straight. I mean, Mum. You *know* what she did.'

The ensuing silence was so long that I started to think that Beth must have ended the call. 'Oh, no, really? I'm *sorry*.' Her voice was heavy with regret. 'I'll pay you back when I've got a job. No, I will. Ok. Well. Thanks. Look, I'd better go. I'm at my friends' place, I can't really talk.' There was a strung-out beat, then she said, in a much smaller voice, 'I love you too.'

She must have taken the stairs two at a time, because seconds later the handle was turning. I leapt away from the door and tried to look nonchalant, but as I closed it behind Beth, we both noticed an ear-shaped imprint of hair dye at head height.

'How much of that did you hear?' she said.

'Caught the shouty bit,' I admitted.

She flumped down on to the futon, messing up Kit's carefully laid-out display of miniature tools, but she didn't seem to notice even as he scrambled madly to right them.

'You remember me telling you about my friend Tess?' she asked.

The name conjured a slowly downloading picture of Beth in her tequila-girl get-up. 'How could I forget?'

'Mm.' She dropped her gaze. 'Well, I possibly slashed the tyres on her car.'

'Beth!' I said, although I could imagine it all too easily. Half admiring, half shocked.

'I don't know what came over me,' she said. 'I saw her pikey little Vauxhall Nova in the Morrison's car park and then next thing I knew I was in the shop buying a little knife. Stuck it straight through the rubber and they just went pffffft.' Her cheeks hollowed as she mimed the deflation.

'Wow,' I said. I looked to gauge Kit's reaction but he was busy rearranging his computer repair tools into height order in their clear plastic box. 'How'd they know it was you? CCTV?'

Beth shook her head. 'She'd given the car to her mum. How was I supposed to know? She came out with her trolley as I was doing the last one. I ran off but she knew it was me; I mean she's known me for ever. She's just been round to my mum's asking her to pay for the repairs. My parents have gone mental.'

'Jesus, Beth.'

'I know, I *know*,' she said. The awkward laugh we shared was one of disbelief, but also seemed to acknowledge that one day there would be a funny side. The last of Kit's humour seemed to drain from him as he sealed his toolbox with a click and pointedly went to put it out of harm's way in the bedroom. Beth lowered her voice. 'I don't expect much from men, any more,' she said so quietly I had to lean in close. 'I pretty much expect them to fuck me over. But to have a *girlfriend* betray you. It's on a completely different level.'

I imagined how hurt I would be if Ling had sold me out like that and knew it would destroy me. I would want to slash her tyres, too. I would want to do worse.

'It's just as well you've got me, then, isn't it?'

I could see from the way Beth's face shone that it was the right thing to say.

34

LAURA
30 July 2000

In July I landed my dream job, a year ahead of target, as a fund-raiser for a children's charity. It was the post that set me on my current career path, although I wonder now how I had the energy to perform at work. It was a sweltering summer, and Kit and I spent most of it on the Underground; if we weren't commuting we were going up to Turnpike Lane to look after Juno (me) or deal with Mac (Kit). We took twice-daily showers, some of our dark clothes were ruined by salty sweat rings, and when we blew our noses the tissues turned black.

Beth was a regular presence in our flat. At her parents' home in Nottinghamshire, she was a recluse twice over, the shame of being outed as the Jamie Balcombe girl compounded by her deft way with a knife, but on her fortnightly or so visits to London she showed no sign of wanting to explore the city, preferring to hunker down indoors. Apart from our meeting on Lizard Point, I had barely seen her in daylight. We didn't discuss that she would always stay over; it was one of many things that went unsaid.

Jamie wrote to me twice a week; letters came without fail on Tuesdays and Fridays. The gist was always the same – retract your statement now, before things get even worse for you – but he would always tell me about other developments. He'd built up quite a network of supporters, apparently; other 'wrongly accused' men were writing to him in droves. There was a men's rights movement concerned about what they perceived to be a trend for women 'crying rape'. And – I could see his glee in the strokes of his pen – women wrote to him, too. Women he called '*real*' rape victims, girls who'd been fucked by their stepfathers, drugged in nightclubs, gang-raped at knifepoint. Jamie never missed a chance to compare these 'brave ladies' with my own cowardice. I tore the letters up and distributed them between various litter bins between our house and Clapham Common Tube. I began to wish that the man with the razorblade embedded in a toothbrush would pay a visit to Jamie.

Ling had finally been prescribed the medication that would haul her out of her depression, and had decided that the time had come to stage an intervention for Mac. This US import is part of the vocabulary now, but this was the first time I ever heard of one, and she had briefed us on the process. All of us – me, Kit, Ling's parents, Adele, plus an addiction counsellor Ling had found – were to turn up at their flat and tell him one by one that things needed to change. The idea was that on seeing this critical mass of concern from everyone who mattered to him, he would march from there straight into rehab. Kit didn't for a moment believe it would work, but he came along anyway, all of us crammed into their tiny basement flat, where there were so few chairs we had to take it in turns to sit down.

There was a snag in Ling's plan. For the intervention to work, Mac had to be present. And, as the main symptom of his addiction was disappearing on benders that took him away from home for nights on end, the odds were against her. Waiting for something horrible to happen is almost more draining than it actually happening, and after eight hours during which Mac didn't answer his phone, let alone come home, we called it a day. The train was like an oven, we got stuck between stations for twenty minutes and we both hated everything, including each other, by the time we surfaced.

'Oh, for fuck's sake,' said Kit as we rounded the corner on to Clapham Common Southside. 'This is literally the last thing I need right now.'

I followed his gaze; a pair of scuffed silver trainers and slender white ankles stuck out from our doorstep. 'Did you know she was going to be here?'

I returned his hostility. 'Of course I bloody didn't.'

I had nothing left for anyone but Kit, and not much for him. Our day must have shown on our faces but Beth seemed impervious to our exhaustion.

'I can't stay long,' she said, leading us up our own stairs. The violet in her hair had faded already; pastel hooks broke up the mess of black curls. Her hand trailed along the wallpaper; at one point she stopped on the landing to examine a patch of peeling green. 'I've got to catch the last train home. I'll be cutting it fine now; I was expecting you back earlier.' Kit and I exchanged a puzzled look; Beth never went home. Once she was here, it was accepted that it was for the night.

'Home?' said Kit, raising his thumbs then hiding them in his fists as Beth turned back, so he looked like he was sparring his way up the stairs.

'Stop it,' I mouthed, but I was smiling.

'Yes,' said Beth. 'But I thought I'd give you the good news in person.'

It must be the appeal, I thought; they must have dropped it.

'Go on,' I said. We were on the top flight of stairs now and Kit was fumbling for the keys in his jeans pocket.

'I'm moving to London!' Beth beamed.

I almost missed the next step. Kit froze with the key in the lock, and I knew he was thinking the same thing as me: she reckons she's moving in here.

'Tell me more,' I said cautiously.

'Well, I told you about the atmosphere at home. I'm virtually under house arrest. So I've been house-hunting and job-hunting,' she said proudly. 'I've got a job at Snappy Snaps, working in the darkroom. Pay's shit, but it's working with photographs. I had a look and I can't afford anything round here, but I've got myself into a bedsit in Crystal Palace. I can see much more of you now.'

'That's brilliant,' I said, although my delight was in knowing that we would see less of each other now; or rather, we could scale the friendship back to something less intense, more normal. We'd be able to meet for a coffee for an hour, or dinner for the evening, then go back to our homes.

'So. I just got you one last thing, to say thank you for having me. I know you said I didn't have to keep giving you stuff, but this is the last one, I promise.' Her smile was shy and child-like. 'Please don't tell me I shouldn't have.'

The present she handed me had been carefully gift-wrapped in William Morris paper and tied with a real ribbon. It looked like a book, but I opened the paper to find a slim pine picture frame and in it was—

'Bloody hell,' I said.

It was a photograph of me and Kit, clearly taken from the doorway of our bedroom. When? Early one morning, at any rate, while we were still asleep. A dawn light streamed in between the slats of the bamboo blind and painted tiger stripes on our skin. The covers were half off, almost as though they had been pulled, and we were exposed from our waists upwards. I was lying on my back while Kit was on his side, half curled around me. His arm was slung across my breasts, his fist clutching at my hair as a baby might its comforter. Instinctively I crossed my arms over my body. Kit sucked in his breath; I could feel the anger coming off him in waves.

'What do you think?' Our silence was protracted and Beth's eyes were starting to lose their shine.

I couldn't use the word *violation* knowing what she had been through, but there was no other word to describe how it felt to see myself like that.

'It's very . . . intimate,' I said, at last.

'I know!' she said. 'You can't plan photos like that. But the door was open when I went to the loo, and I couldn't help but see you and the light was just . . .' She held up her fingers in the OK sign. 'And your camera was sitting on the side, Kit.' His mouth set in a furious line; he had a stronger reaction to the thought of her touching his precious camera than he'd had at the sight of the photograph. Beth continued, heedless. 'And I thought about that other photo that your friend took, and what you said about not being aware of it, and I couldn't resist. I went to the one-hour photo place on the common today.' Her voice weakened a little with every sentence. 'I used a really slow shutter speed. The lens is . . .' She shook her head, her voice diminished to a whisper. 'You don't like it.' She scanned our faces and read them wrong. 'I'll replace the film,' she said, missing the point by miles. 'It was only from Boots. Sorry, Kit, I didn't even think you'd notice it was gone.'

'It's fine,' I said, but Kit's silence was louder.

Beth hit herself hard in the forehead with her palm. 'I thought you'd get it, I thought you'd *love* it.'

'I do love it,' I said, catching her wrist in case she struck herself again. I had forgotten how soft her skin was. 'It's just unexpected, that's all.'

She wriggled free of my grip. 'I can be a bit full-on,' she said, addressing the window. It had the ring of something she'd been told about herself more than once. 'I've misjudged it. I'm sorry.'

Later, after an awkward goodbye, Kit and I waited in silence for the street door to slam behind her, unsure whether to laugh or scream.

Kit held the framed picture at arm's length. 'How could she *possibly* think that was ok, on any level?'

'Well . . . I suppose her judgement's off, after what she's been through.' For all we knew she had always been this direct, always had such fluid boundaries; or perhaps it was a reaction to her ordeal, that she had been stripped back so far that she had decided to leave the nerves exposed in the hope she might eventually become desensitised. This was Beth's trauma. If this was her way of processing it, who were we to judge?

'What if I'd had my knob hanging out?' said Kit.

I stood next to him and looked at the photograph properly; it was easier without the weight of Beth's expectation.

'We look good, though.' I smiled. 'If you take away the *massively creepy* nature of it, it's actually quite beautiful. I didn't know you slept with my hair in your hand like that.'

'Neither did I,' he said, softening. He smoothed down the length of my ponytail.

'We can look at this when we're old and saggy,' I said. 'Remember what we used to be like. I know it's not exactly one to have on the mantelpiece, but I'd like to keep it.'

We put it face down in my bedside drawer. It was one of the things that we lost when we went into hiding, and I don't know where it is now.

Looking back, Kit was right, and I should have been firm but fair at the first opportunity, that time she'd come to see me at work. I shouldn't have let her into our lives. I should have kept the relationship *proper*, to use an old-fashioned word.

That night in bed, I couldn't sleep. I was too aware of what I looked like; too aware of being naked. Only when I got out of bed and put some nightclothes on did I finally relax. I had weird dreams – a sense of being watched, of a figure in the doorway – that I attributed to Beth's photograph. I didn't yet realise that my unease was about something more than her trespass. It was the first stirrings of something too big, deep and dark to shove in a drawer or a bin in the street. Something I was not yet ready to name.

35

LAURA
20 March 2015

First contact will be at 8.20. The local radio says that the capital will be 'plunged into rush-hour darkness.' I could watch the partial eclipse, as it will be over London, from the study, but even with the window open to the white sky it feels wrong to observe it from inside a building.

I braid my hair into a skinny plait that hangs to my waist, then wind it into a bun. Hiding my hair is second nature whenever I leave the house. I'm like a woman whose religion compels her to cover her hair in public; only my husband really sees it loose. The alternative was to cut it, or stop colouring it, and she has taken enough of my soul already. I'm not giving her that.

I decide to take a walk around the neighbourhood, to soak up some of the atmosphere, but there is none. Up and down the length of Green Lanes, cars beep and lorries unload like normal. The snooker hall is shuttered at this time of the morning. I stopped going there a few months ago, when my belly got too big for the table. There are only so many trick shots you can take with a cue behind your back. The only clue an eclipse is imminent is not in the stars but in the gutter; a discarded *Metro* newspaper is open at the headline: TAKING SELFIE WITH THE SUN COULD BLIND YOU, SCIENTISTS WARN. There's nothing dramatic overhead but a weird violet light that I might have put down to a coming storm if I noticed it at all. It's only on Duckett's Common, where a handful of people dangle goggles from their hands, that you'd know anything special was happening. I stand, feet planted firmly, both hands on my bump. The capital is not plunged into darkness; it's barely plunged into dusk. Even the babies share my sense of anti-climax, sleeping through the whole thing.

Back at home, there's a Bone/Bean bag on my doorstep and a note inside telling me that Juno and Piper came round with Mac to deliver my morning liquid diet. My heart dips for having missed them. The coffee is still warm enough to drink, but I pour the bone broth down the sink and fill the fridge with the juices. The computer upstairs is calling me – I have a reference to write – but I know I won't settle to anything today. I keep checking my phone, waiting to hear from Kit and forcing myself to stay calm when I don't. He's told me half a dozen times that he will probably be out of signal all morning. I swipe to my messages and look again at his clean-shaven photograph; a sweet gesture meant to placate me but all it does is remind me why I was so angry with him in the first place.

I turn on the TV. The woman on the BBC says that conditions in Svalbard were perfect but that Tórshavn was a failure. I feel sorry for Kit but there's a surging anger inside me too; what a waste. All this stress, all this churning up the past, all this money spent and one of the worst rows we've ever had. Without a clear eclipse to justify it all, I feel cheated. It's all been for nothing.

The phone rings. 'What a total bloody let-down,' says Dad.

'It was supposed to be good up North,' I say.

'It got very slightly darker,' he concedes. 'How'd Kit get on?'

I realise that Kit and I haven't spoken since our argument. What if that was our last conversation? There's a sudden swelling inside me, like someone's inflating a balloon inside my mouth. I want to tell him how stupid Kit's been, how he's told Beth where to find him, but I can't, because it's all down to me.

'Daddy?' I say, to our mutual surprise.

'You all right, darling?' The concern in Dad's voice pops the balloon. If I break even a little bit, the whole thing will come spilling out. 'Fine,' I reply. 'Bit of heartburn. I haven't spoken to him yet, but it doesn't look good,' I manage.

'Shame,' he says distractedly. 'Listen: one across; the wrong thing for the right reason. Six, five, three.'

I don't have a clue. The analytical part of my brain has been temporarily subsumed by the one responsible for paranoia and foreboding.

'I'll have to think about it,' I say.

'Make sure you do. It's got the first letters of five other words locked up in it.' He clears his throat. 'How are my grandchildren doing?'

'Good, I think.' Right on cue, a little foot pushes at the wall of my stomach; I bring my hand down as if to tickle its toes. We arrange a day for Dad to visit next weekend; there's a new ocakbasi restaurant on the corner I want to take him to. By the time I've briefed him on this afternoon's scan, and heard the latest developments in his doomed election campaign, it's time for me to go to hospital. My maternity notes are neatly bound and in a box file next to my computer.

At eleven o'clock, I check my phone but there's still no contact from Kit. My arms start to prickle, even though he told me not to expect a call before lunchtime. I pull my sleeves down over my wrists and slide my maternity file into my bag.

The solution to one across hits me then. The wrong thing for the right reason, six, five four. Little white lie.

36

LAURA
29 August 2000

Kit's open, glowing laptop was usually a marker of his presence in the house, but it had been shut for five days – a record – and the pile of essays he had brought home to mark over a week ago lay by the front door where he had dropped them. After a string of failed interventions on our part, Mac had staged his own, of sorts, getting himself sectioned after crossing live railway tracks between two crack houses in Tottenham. He was currently undergoing treatment for minor burns, cuts and grazes and major addiction in the secure wing of the North Middlesex Hospital. Today, ten days into the treatment, was the first time Kit had been allowed to visit.

While he was out, Beth and I had worked our way through a crate of French beers. She had turned up unannounced, but I'd withheld the usual invitation to eat with us. I'd prioritised her for long enough; it was time to put Kit first again for a while. Although I was never going to turn her away, like he wanted, since the photograph I felt the need to keep her at a distance.

The three of us were dealing with the incident in a mature and responsible way, by pretending it had never happened.

If Kit had let me explain his state to Beth, perhaps she would have understood, and given us the space we so desperately needed. But, fiercely protective of his brother, he begged me not to let her know, and so she kept coming blithely to Clapham and I had neither the heart nor the words to send her away.

It was Tuesday, and in this morning's letter, Jamie had wondered whether to address next week's plea directly to Kit. I reached for another beer.

'You've hardly used those candles I bought you,' said Beth. The three glass lights were still lined up on the shelf, the wax in the middle one pooled a centimetre lower than the others. The truth was that I'd rather gone off the scent.

'I'm saving them for a special occasion,' I lied. 'It's more of a winter thing, burning candles. The evenings are still light, the window's always open. Do you want one, for your place? It might take away the smell of damp.'

'Only thing to take away that smell'd be if someone burned the whole shitpile to the ground,' she said glumly, her fingernail edging the label off her beer. 'Anyway, I bought them for *you*.'

I lit one, but placed it on the balcony, where most of the scent would be carried away.

Four storeys below, the street door slammed.

'We're in here!' I shot a warning down the stairs; the *we* rather than the *here* clearly being the key information. By the time Kit came into the sitting room, he'd dug up a grin from somewhere.

'Been anywhere exciting?' Beth asked him. He let the smile slip for a moment, just to shoot me a look that said *We do not talk about my brother in front of her*. I gave a tiny nod.

'Not particularly,' he said. He took his Swiss Army knife from the kitchen drawer and levered the top off a bottle, bringing it to his lips as the foam burst over the neck. He drained it in a series of gulps, then opened another – I'd never seen him do that before – before flopping on to the futon and glowering at the wall opposite. I wished Beth wouldn't stare at him, or would at least try to disguise her fascinated concern. The little flat swelled with its various secrets; it reminded me of the witness room at Truro Crown Court. Kit, who had long dismissed the theory of twin telepathy and phantom pain, had sweat on his upper lip and kept shifting like something was cramping in his belly, for all the world as though he was the one going through cold turkey.

'Are you all right, Kit?' asked Beth. 'You don't look very well.'

'I'm fine,' he said robotically.

'I'll put the telly on, shall I?' I said, aiming the remote. A news channel was showing a story about a proposed bypass tunnel to take traffic away from Stonehenge.

'It'd be a shame if you couldn't see it out of the window,' I said. 'I love that first glimpse of it on the hill.'

Kit grunted an acknowledgement that I'd spoken.

'They need to sort that road out, though,' said Beth. 'It's *always* congested; it adds about two hours on the journey time to Glastonbury. It's bad enough when there isn't an accident, but when there's a crash it's literally impassable. There was a pile-up there the day I went to the Lizard, about half a mile in front of the car I was in. We were stuck in traffic for about five hours. It was only a little Ford Fiesta, my knees were in agony not being able to stretch out. All the wreckage was still smoking when we went past it.' I might never have understood the significance of her words if it hadn't been for Beth's blush. I've still never seen a high colour like it: hot pink patches started to form on her white neck and crept steadily up her face, like someone was pouring the blood into her from a jug. 'Anyway, so, I just think a tunnel would be best.' The blush had evaporated and she spoke with an almost aggressive brightness. 'Stop them concreting over any more fields.'

The next item was a report about flaws in the Inland Revenue's new online filing system. Beth, rather unexpectedly, had plenty to say about this, too; apparently her solicitor was calling it all the names under the sun, he needed to get with the times . . . I tried to tune her out and pick apart what she'd said about Stonehenge but her prattle was relentless, like a toddler's. It was almost as though she was deliberately occluding thought. Only when she disappeared to the bathroom did the penny drop. I was on the verge of asking Kit if he'd noticed her go red when it came to me;

my admittedly theatrical gasp was loud enough to pull him out of his mood.

'What?' he asked.

'Ok, this might be nothing,' I said, even as momentum gathered inside. 'Just hear me out, though. The crash Beth was talking about happened the day *before* I came down.' I paused to let the weight of it land on Kit, but his face was blank. 'The verges were still covered in debris,' I said. 'The coach driver told us that they couldn't clear it until all the eclipse traffic had died down. But they said in court, didn't they, that she only arrived the day before the eclipse?' Something else occurred to me. 'And they said she'd travelled by coach, because I remember thinking she must have been on the first one of the day, if she wasn't on mine. She just said, didn't she, that she was in a car.'

Kit screwed his face up. 'Yeah . . .' he said eventually. 'I'm pretty sure you're right.' Of course I was right. I could have sat a test on that trial and aced it. He shrugged. 'Polglase must have made a mistake.'

I shook my head. 'Fiona Price would've been all over it. So would the judge.'

'Then Beth's probably mistaken.' I couldn't believe how dismissive he was being. 'She must have just seen what *you* saw. All the debris piled up at the side of the road.'

Beth was humming in the toilet; I dropped my voice even further. 'I don't think so. Kit, I think she's lied. Even if she's wrong about the day, what about the car thing?' The toilet flushed noisily. 'I'm going to ask her.'

'Oh, *God*,' said Kit. Finally I had his attention. 'Don't set her off. Can we not just have a quiet evening in?' But I couldn't let it go. My lie had been built on the conviction that Beth was telling the whole truth. If she hadn't been, what did that mean for me? I went very cold. She came back, fluffing out her hair. I cleared my throat.

'Beth, I was just wondering – and it's probably my mistake, but you know what you just said about seeing that accident?' She didn't blush this time but went white; even her lips lost their colour and I realised even before I spoke that I was on to something. That if I *set her off*, to use Kit's words, this was going to change things. I had to know, but it took courage to keep talking. 'It's, the thing is, it's just that they said in court that you'd said you came down on the Wednesday, like me, and that you were in a coach, but you just said you were in a car. I just wondered where the mistake was. If it was your mistake or the court's. Or . . .' I was in so deep I might as well keep going, 'If it was a mistake at all.'

She folded her arms. 'Oh, I'm sorry, I didn't realise I was on trial. *Again.*'

'No, don't be like that, I don't mean it like that.' Didn't I? Suddenly I wasn't sure. I took a deep breath. 'Say Jamie's case *does* come to a retrial, that's a discrepancy that could trip you up. I'm trying to help you out here.' I looked at Kit for back-up but he was looking at the floor, like he wished the pair of us would vanish in a puff of smoke.

'Ok.' Beth leaned back against the wall and closed her eyes. 'If you must know, I lied about the day I came down because I didn't want them to know how I got there. I didn't get one of the special coaches. I hitched.'

'You *hitched*?' Incredulity made me temporarily blind to how close I was scooting to my own lie.

Kit's eyes followed our conversation back and forth.

'Yes.' Beth opened her eyes; her chin had a defensive tilt to it. 'Hitch-hiked. I stuck out my thumb and made a sign and eventually people picked me up. I had an old couple take me as far as the Fleet services, then a couple of girls took me to Helston, then a bloke in a Beetle took me right on to the site. I'm still shitting myself that Jamie's defence might find one of them. I kept expecting to see him every morning in court. I reckon their lawyers have got one of them now.'

I was simultaneously relieved and bewildered. 'I don't get what the big deal is, how and when you got there.'

'Do you really not?' I shook my head. Beth sat down next to me on the futon and took my hand. I saw a faint black down on her upper lip that I hadn't noticed before. 'Because it was obvious to me, even in the first few minutes.' She nodded at my bookshelves. 'You of all people, with your Germaine Greer and your Camille Paglia, should know. Even *I* read the newspapers, I knew what they did to rape victims. If they knew I'd hitched down, they'd have said I had a history of risk-taking behaviour, or worse. I already knew that it was my word against his. And so I thought I'd take away anything that could be construed as asking for it.'

Speech over, she sank into the cushions and waited for me to say something, but I was still trying to process it all. I had *seen* Beth in the aftermath. She was too traumatised even to tell me her name, let alone think on her feet like this. She took my confusion for doubt.

'You see?' Beth threw up her hands. 'This is why. I knew they'd judge me. You are too.' She was on her feet, shoving her things into her bag. 'I'll leave you guys to it.'

'Beth, please, don't go like this,' I said. Kit glared at me; this was exactly what he wanted: space, until the next legal step, and he didn't care if this was the price. But I could no more let this go than he could drop his twin. I was part of this story and I needed to understand all of it.

Beth swept her stuff off our mantelpiece into her open handbag – keys, phone, purse – in a bad-tempered flounce, then bent to tie her shoelaces. 'I'll be in touch if there's any news about the appeal,' she said.

'Stay for a drink,' I said desperately, 'Let's talk this through.'

'Talk this through?' she said. 'That's a fucking laugh. Do you have any idea what it's like, spending all this time with you both and having to constantly bite my tongue?'

My blood banged against my skin. She's going to say it, I thought in horror. She's known all along that I lied for her in court and she's going to tell Kit now. I could tell from his creased face that I hadn't hidden my horror in time. I got the feeling that he was somehow reading my mind; that he knew how big this could be. I knew that my script read, 'What's that supposed to mean?' but I couldn't risk the answer.

'Fuck the pair of you,' said Beth, and she was gone, bag jangling and hair flying.

After the front door slammed, Kit and I were left blinking at each other in stunned confusion and, in my case, a reprieve that could only be short-lived. I needed to go after her. To beg her to keep biting her tongue for my sake. Kit was the first to move, leaning over the balcony.

'Where's she gone?' I asked him.

'On to the common, through the trees.'

I flew down the stairs and across the grass to the bus stop, only realising on my way that I didn't know whether she was heading to Crystal Palace or back up to Nottinghamshire. I checked the bus stops on both sides of the road but I'd missed her. I circled the common twice before giving up.

When I got back, much later, there were three empty bottles already lined up in the recycling. Kit got up to fetch another beer from the fridge, opened it, and gave it to me before getting his next one. Now that Mac was drying out, Kit seemed to be drinking for two.

'What the bloody hell happened there?' he said.

'I've clearly hit a nerve,' I said. 'She's hiding something, Kit.'

'She might not be. Self-preservation is an amazing thing. And you said yourself, people react weirdly to trauma; there's no such thing as normal.'

'Mmm. I should talk to her tonight, smooth it out. I've really upset her.' My eyes were dry from staring across polluted streets; I balled my fists into the sockets. 'It's just as well we can't afford a car, isn't it?' He didn't get the reference. 'I mean, she might have a go at the tyres . . .' My joke fell flat.

'Well, I hope she meant it, about backing off till the appeal,' said Kit. I leaned into his chest; my ear found his heartbeat. His arm was heavy on my shoulder. 'It'll be good for her. She's too dependent on you. You can't carry her forever. And . . .' Kit pulled out of our hug, took a huge gulp of air, then blew it out for twice the count, like a yoga exercise. '*I* need you. I can't . . . Mac . . . I . . . he looks like he's going to die. I didn't know what to say to him. I'm out of my depth.' He inhaled the next sentence. 'I'm fucking – all this shit . . . I'm *drowning*, Laura.' It was the first time I'd seen him cry since Lachlan died, and entirely different. Grief had drawn from Kit a slow, steady sobbing but this was a series of explosions, each wordless shout more powerful than the last; his tears were heavy and copious. I tried to put my arms around him but he swatted me away, although when I put a hand on his back he let it stay. He dropped his head down into his chest, his body a tight curl, making me aware of an almost prim straightness to my own back. I kept my palm flat against his ribs, feeling his lungs punch against them, until he had cried himself out.

The next day was a Saturday; we lay in for hours, the slats of shade a crawling sundial on our skin. Confident that Jamie wouldn't write today, I held back from checking the doormat; I didn't want Kit to notice that I was always rushing down to fetch the post. It was lunchtime when at last he got round to it.

The scream that ricocheted up the stairwell was as high and girlish as last night's tears had been masculine. I only thought for a tenth of a second it was one of Jamie's letters; the noise was too primal.

I met Kit on the third-floor landing, a bloodied shard of glass in his left hand. He'd gone pale; freckles I hadn't noticed in years stood out on the bridge of his nose.

'It looks worse than it is,' he said, unconvincingly. 'Some bloody Friday night pisshead thinking it's funny to push this through the door. We're getting one of those wire baskets for the letterbox.'

The timer on the light went out and for a moment we were in pitch darkness. I felt my way to the top of the landing and hit the switch. Kit had shifted, his left foot upturned to display a three-inch gash along his sole.

'When did you last have a tetanus booster?' I asked.

'Last year,' he said.

'Let me have a look, see if we need to take you to hospital.' He hopped his way to the futon. I laid his foot across my lap and trained the anglepoise on his sole while I checked him for splinters. The cut was long but shallow, and already beginning to clot. There was one more sliver embedded inside his arch. 'I think your footballing career is over,' I said, as I homed in on it with my tweezers. '*There*.'

'Wheurgh,' he said. When I squinted to examine the glass, I saw and smelt the evidence at the same time. A smear on the fragment of palest pink wax; the trace scent of Blood Roses. My eyes travelled to the mantelpiece. The middle candle, the one I'd started to burn, was missing. I remembered the way she'd swept all her stuff into the bag. Had the candle been caught up or deliberately stolen? I pictured her, hanging around until our light went off before feeding broken glass through our letterbox, and reached for the support of the armrest. Was this vicious act the flipside of the adoration she'd shown so far? And why? Because we had challenged her story? Or was she just lashing out because of the stress of the appeal being dragged out for so long?

'This is my candle,' I said to Kit. I held the glass out in my hand. Not we, I realised with horror. Me. She knew I was always the first one downstairs. Me. 'Someone's smashed up the candle and then posted the bits through the letterbox.'

I didn't need to name her.

Kit blanched even further. 'Why would she do that? That would be *mad*.'

That night, when Kit was asleep, I went on his laptop and read every word of jamiebalcomeisinnocent.co.uk. with a mounting sense of discomfort. In the last twenty-four hours Beth had shown herself to be duplicitous and vicious, and my previous certainty had been cut adrift. I read Antonia's statement over and over again and Fiona Price's bullseye rang in my ears. 'You got carried away with the drama, didn't you?' What if I had not got carried away with Beth's drama, but she with mine? Had Beth *gone* mad because Jamie raped her? Or had she said he raped her – or gone along with my assertion that he'd done it – because she was *already* mad?

Until that day I'd been thinking about the campaign in binary terms: either Jamie was in denial, or he was calling the world's bluff. But now a third possibility was inescapable, and one in which I was horrifically culpable. What if Jamie was innocent?

37

KIT
20 March 2015

Rain cover off. Rain cover on. Lens filter off. Lens filter on. Battery removed, checked and reinserted. Strap adjusted.

'Would you stop fucking fiddling with your camera,' says Richard.

'Sorry.' I set my hands back in my lap and resist the urge to drum my fingers on the window of the bus.

'Sorry mate, I didn't mean to snap,' says Richard. 'It's just, you know.' He gestures out of the window. The atmosphere in the bus as it wheezes up Húsareyn resembles the sky outside: dark and rumbling. The clouds are fast-moving but thick. Sheep dot the fields; here and there, black rocks jut through the tufty grass and heather. There's a flash of colour on the horizon; three pink coaches, beetling up the neighbouring mountain. Instinctively I throw my hood up.

As we park at a tilt on a stony verge, I wonder what it's like in London. It would be ironic if Laura gets a clear view of the partial while I'm stuck clouded out of totality.

We settle on the hillside, the toytown harbour spread below. It is relative wilderness here. There is enough space for all the parties to spread out. Rather than the huge crowd I'd feared, we are in clusters. Of the few lone figures yomping out across the uneven ground, none of them is her. I'm sure that even after all this time

I'd be able to recognise her, even if only by the roll in her walk, the curve of her, so different to Laura.

I set up my camera tripod and fiddle with the settings, squatting to squint through the viewfinder.

'I'm surprised you can see anything out of that hood,' says Richard.

He's right, it is obscuring my vision. But I'm not taking any chances.

Ten minutes before first contact, the sun peeps out from behind a lacework of clouds and then, as if startled by our cheering, retreats. 8.29, first contact, and it still hides. I keep my Mylar glasses on, but it's only a gesture to potential. As the moon chews its way through the sun, only occasional glimpses of the shapes in the sky are available, and as totality approaches, the clouds thicken until the sun's position above is hardly discernible.

My eyes keep travelling to a loose rock at my feet, sized halfway between a fist and a human head, and I think: this will do. If she looms from nowhere, this will do me just fine. I'm horrified at my own thought process. This is not me, I tell myself, this is not me; and then, she's not here. She's not here.

'Who's not here?' says Richard. I didn't realise I'd spoken out loud.

'Is that a break in the cloud?' I deflect, looking up at the impenetrable sky.

Disappointment momentarily crowds out thoughts of Beth, and it's almost a pleasure to feel a different negative emotion. I decide instead to try to observe the other phenomena that come with the eclipse, the things you miss when you're transfixed by the sun itself. I've always been so busy looking up that I've never observed the flowers closing, for example. But there's nothing underfoot but rocks, scrubby vegetation and sheep shit.

Then the darkness comes. Without the countdown in the sky, it's pure and instant. All the street lights in the town below come on, so quickly it's like sparks igniting. Now, in the darkness, disappointment is replaced by the familiar thrill of totality. But this time it's different. Suddenly, unexpectedly, fear pours down through the hole in the sky and I am overwhelmed with a little child's terror of the night, only my monster is what Beth represents; she embodies all that I stand to lose. The air all around seems thick and sound muffled; at one point, I sense something behind me, and whip around from the camera, but there's nothing there.

I keep the lens trained on the sky in the increasingly vain hope that something might happen, but I can't focus on the clouds, only on the darkness around me and behind me. The mountainside holds its breath.

Maybe ten seconds after third contact, we see a crescent sun, but I can only release the shutter a handful of times before the clouds return. Then the black water in the bay is silver again, and it's over. The post-eclipse world is flat and ordinary. There is no Beth, and even if there were, I'm not powerless against her. I see that now. The fears of a few seconds ago seem unfounded, as nightmares always do when the light comes back on.

38

LAURA
4 September 2000

'I've got half a mind to report her for assault,' said Kit, gingerly easing his foot into a shoe for the first time since his injury. It had been meant for me, and I was in no doubt why. I had never said that it was probably time to put it all behind her now, or asked her what did she expect if she went to a festival on her own, or said that she must have done something to make Jamie think she wanted it, or suggested that on some level she must have enjoyed it. Other people said those things to Beth and worse. They said these things to her in court, then in the press, then on the internet, on the streets of her home town. People said terrible things to win the case or in spite or even out of love, but until that one night, I never had. From Beth's point of view, it was Tess all over again. I'd drawn her in, only to throw her out.

'Don't do that!' I begged Kit. 'She's disturbed. She's a *victim*. We can't turn on her.' He made the old-man's harrumphing noise he reserved for the times when he knew I was right but he didn't want to take me on. I forced myself to speak more gently. 'I'm not ruling it out. But I'd rather talk it through with her.'

'And have you?'

He knew perfectly well I hadn't picked up the phone. I was still going over it in my head. The situation with Beth, once so clear, had been muddied, and I wanted to let the brackish waters of my mind settle before I talked to her.

'You can't throw people on the scrap heap just because they fuck up once,' I countered.

'It's hardly a one-off, is it? Turning up at all hours, bringing us over-the-top presents like a bloody cat bringing in mice? You heard what she did to her friend's car. And that's before we even get on to the naked photographs. She's bad news, Laura.'

Despite everything, I still felt protective towards her. 'Anyone would think you'd forgotten how we met her.'

His nostrils flared. 'There's hardly any danger of *that*.'

We stared at each other, our tempers on a hair-trigger. Kit backed down first, as I knew he would if I waited for long enough; it was rare I exploited his ingrained acquiescence but I made no bones about it now. 'Look, I really don't want to fall out with you over her,' he said. He opened his arms. I leaned against him but let my hands hang heavy by my sides.

I spent two more days trying to think of an opener for my talk with Beth. Cold-calling companies at work was nothing compared to this conversation. Beth's skin was a thick hide in places, little more than a membrane in others. You never knew how she was going to react.

In the event, she rang us. As soon as I knew who it was I put her on speakerphone. Kit came to stand next to me in front of the phone, arms folded, frowning at the floor.

'They threw out the appeal.' Beth's voice bounced off the walls of our little flat, the machine's distortion making it impossible to judge her tone. 'We were right all along. It was eyewitnesses from the campfire. The judges said they weren't relevant to the issue of consent. All that money spent and he still can't escape justice.'

'That's brilliant,' I said. If she noticed the detachment in my voice, it didn't seem to bother her.

'Can we go out to dinner to celebrate? My treat. To say thanks for everything you've done for me.'

I didn't answer quickly enough.

'Laura?'

'I'm here.' I drew a fortifying breath. 'It's just, dinner? After what happened last time. We hardly parted on the best of terms.'

Kit flexed his stockinged foot, an unintentional reminder of what she'd done. Beth didn't know he was listening in, I registered. The old panic crept in that she might choose now to broach my lie in court.

'Yeah, well, look . . .' I recognised the strain of control in her voice. 'You said things, I said things. I mean, I won't pretend it didn't hurt. But I can forgive and forget if you can.'

I felt the weight of Kit's foot in my lap again, blood oozing from torn skin. 'It's hardly the same thing,' I said. 'It's hardly in the same *league*. You can't just lash out like that!'

Beth paused before speaking. 'Laura, emotions run high when you've been through what I have.' It was the first time she'd directly used her ordeal against me; at least I stopped myself from saying she couldn't use that as an excuse. 'I don't understand why you're being so weird about it. What do you want me to say?'

'Sorry would be a start,' I said.

'*Me* apologise to *you*?' Her breathing went heavy, then stopped altogether.

'She's hung up on me!' I said to Kit. I replaced the receiver with belated gentleness. 'Why would she do that?'

'I don't know,' said Kit cautiously. 'She does seem to sort of shut down when you challenge her.'

'I'm ringing her back,' I said.

Kit gently put his hand over the receiver.

'Maybe you should calm down a bit first,' he said. 'Your hands are shaking.'

I was shaking all over. The only thing worse than a confrontation is having it snatched away from you. But he was right; if I spoke to her now, I'd say something I regretted.

'How has this all gone wrong so quickly?' I said. 'I thought she was my friend. I didn't even know her.'

To his credit, he didn't say – he has never said – I told you so.

* * *

No room should be more familiar in the dark than one's own bedroom but suddenly it was as strange as a hotel room in the middle of the night. The smoke was a spur in my throat and pins in my eyes. I managed to pull on a long T-shirt and the knickers that were still on the floor where I'd left them.

'Kit!' I shook his shoulders. 'Something's burning.' That was an understatement. The whole flat was on fire. 'Kit, for fuck's sake, wake *up*!' He had never seemed heavier, or more asleep, and for a few awful darkening seconds, I thought he was dead. 'KIT!' I smacked him full in the face and he coughed himself awake, assessing the situation in seconds. He pulled on his shorts, immediately wide awake, focused.

'It's coming from the staircase,' he said. 'Get on the bed.'

He forced open the window that doubled as a fire escape; it turned the whole flat into a flue, sucking the smoke through our bedroom and bringing with it a huge cascade of orange flame. We both ducked as things escalated next door. There was some kind of explosion; even through the boom we heard the glass shatter in the sitting room. Smoke scraped me out from the inside. 'All our *stuff*,' I gasped. I meant just one thing: the photograph of me and my mum in Greenham Common. The thought of her picture blackening and curling seemed intolerable; preventing it felt worth risking my life. It can't have been more than three seconds but grief is cunning that way. Like smoke, it finds and fills the tiniest spaces. Kit pulled at my T-shirt; it tore as I made for the sitting room, just as the orange dragon of flame bounded up to the bedroom door.

'What the fuck are you doing?' His voice cracked with the effort of shouting.

'I need my mum,' I said. He knew me well enough to infer the thought process behind the words and acted with superhero speed, charging past me and pulling the bedroom door closed. The sound he made as his bare left hand closed over the red-hot door handle was like nothing I ever want to hear again. The scream broke his voice; it cracked halfway through and gave way to ragged breathing. He pushed me with his right hand towards the fire escape. The metal staircase to the street was engulfed with smoke and hot to the touch, so Kit shoved me roughly through to the safety of the rooftops. We climbed, half-dressed and in bare feet, across to where the tiles were cool and tried to make sense of what was happening underneath us.

'I'm sorry,' I wanted to say, but I had no voice even to whisper.

No other floors seemed to be affected but tongues of bright fire streamed from our balcony. The pavement below was dotted with people in nightclothes and dressing gowns. Someone I didn't know said, 'It's ok, they're on the roof!' and then raised their voice to where we were. 'The fire brigade are on their way.'

We could not reply. It took everything we had to breathe. There was a sweet, meaty smell, like barbecued pork, on the air, and I remember looking around me, surprised that you could smell cooking at this hour; surely all the restaurants were closed. It wasn't until I looked down and saw his bubbled flesh that I realised I could smell the seared flesh of Kit's left palm.

Blue lights circled across the common. The ladders went up and the hoses went on. You could hear the sizzling from the pavement. Kit and I sat on the back step of the ambulance, oxygen masks around our necks and blankets over our shoulders. One paramedic treated Kit's hand while the other called the A&E department at Bart's. I was appalled that I had risked both our lives for a photograph.

'Your landlord's got a lot to answer for,' said one of the fire-fighters, helmet off, sweat carving red tributaries on a blackened face. 'I'll bet you anything it's electrics. I can't think of anything else that'd cause a fire like that in a stairwell. You can see where the plaster's come off the wall; that wiring's got to be sixty years old.'

'It's the force as much as the heat,' said the paramedic, gently wrapping bandages around Kit's palm, making him whimper in pain. I'd never seen beads of sweat on anyone's face before, but his forehead was studded with fat droplets that looked solid as set wax. I could hear him grinding his teeth with the effort of not screaming. The firefighter was called away by a colleague. 'If you'd only touched it we'd be talking about a little sting. You must've given that door handle a good yank. Mind you, if you hadn't, you'd be toast now.'

Another firefighter approached us, his gloved hand flat.

'We've found the culprit.' In his palm sat a tiny charred pink disc that I identified in seconds as a Blood Roses candle, or rather its remaining base. Even smashed and charred it held a ghost of its scent. I didn't need to meet Kit's eye to know that he knew it too. 'If it was up to me I'd have these things banned. They're the new cigarettes in terms of starting fires. Well, you won't be doing that again in a hurry, will you?' He leaned down like we were naughty five-year-olds. 'What were you even *doing*, burning it on the stairs? Seriously, what were you thinking?'

'Why did you . . .?' I asked Kit, just as he said, 'Why on earth would you . . .?' Our voices were not our own, our vocal cords tattered by smoke. My own anger was reflected in his face. 'I *didn't*,' he said.

'Well, neither did I,' I said. In the swelling silence we both put together the same puzzle. Mac and Ling aside, there was only one person who knew our flat, one person who knew where the candles were. I remembered with a swoop of horror the day I'd left her alone in our flat. She'd had eight hours to get herself a key cut. It was the only opportunity she'd had.

She'd wanted control from the start.

The firefighter's face took on a completely different expression. 'You're sure neither of you did it.'

'Absolutely not,' we said in unison.

The firefighter nodded the answer to some internal question. 'Right. We'll take this seriously. We'll get forensics to look for signs of forced entry. There's CCTV on the kebab shop next door, it might yield something. You'll have to make statements to the police. You wait here—' as if we had any choice – 'while I go and call a colleague.'

After his departure, we sat there in stunned silence for a while, watching the firefighters traipse in and out of our smoking building.

I finally faced the naivety that had made me lie in court and risk my relationship, risk my *life*. Taking as deep a breath as my charred lungs would allow, I said to Kit, 'It's not arson, is it? It's attempted murder.'

Totality

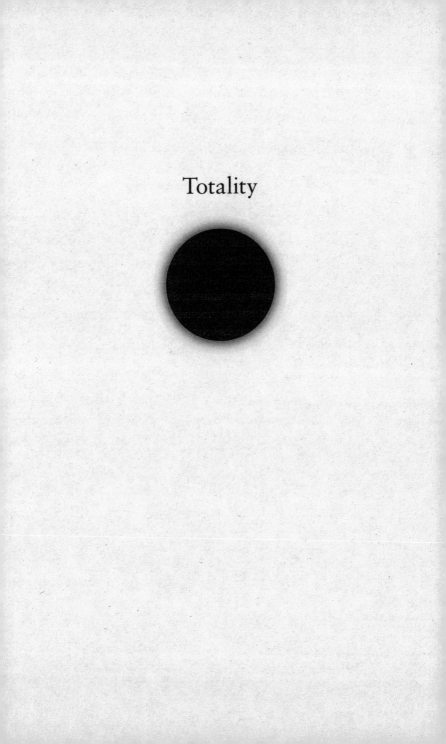

39

LAURA
28 September 2000

Seven days after the fire, I drove Ling's old van round and round Clapham Common, looking for somewhere to park. Kit sat beside me, his left hand still in its big white mitten. On my third loop I finally found a metered space outside our flat. I managed to manoeuvre the van in without scraping the paint off the vehicles at either end, and we fed the meter enough coins to give us two hours.

'I half expect her to be sitting on the doorstep,' I said as I slammed the doors.

The police hadn't been able to question Beth. Although they'd 'had a chat' with her, it wasn't as a suspect. There were no witnesses and the CCTV camera on the side of the kebab shop turned out to be a dummy. The fire brigade had broken the door when they put out the fire, corrupting any evidence of forced entry. I braved Kit's fury – 'What were you *thinking*, leaving a stranger in the house?' – to tell them about the possibility of a copied key, but all the local locksmiths drew a blank. Beth's DNA was all over our house – she had let herself out only hours before the fire – and so forensics would have been meaningless. There was nothing to stop Beth coming after us. My bottomless reservoir of compassion had dried up. Empathy wasn't going to save us when we were choking.

Beth wasn't on the doorstep. Apart from a shiny new lock and hinges, the front door looked like it always did, today accessorised with an empty KFC box and a little sun-dried pool of vomit. A glance upwards at the black feathers around our boarded-up kitchen window told the real story. Fatigue hit me like a wave. I was running off caffeine and panic. I had not slept through the night since the fire, not because I was scared – it was still too early for that – but because there was something I had to do, a cold insistence that peaked in the small hours. Each night I lay awake on the lumpy sofabed in Adele's spare room and vowed I would do it the following day, but days were taken up with hospital appointments, flat viewings and long, circular calls to our insurers and our landlord.

We opened the new lock using a new key. One of the first things Kit had done, almost immediately he was out of the hospital, was to get all our mail redirected to Adele's house; I had transferred my morning post routine wholesale to her doormat. The only letters here in Clapham were the usual flyers and delivery leaflets. The stairwell stank of stale smoke and smut was thick on the walls. I held Kit's good hand.

The carpet still squelched with filthy water. It wasn't the staircase itself that had caught but the walls with their layers of ancient paint and paper. Everything in the sitting room had burned or, in many cases, melted. The television, Kit's camera and laptop were lumps of plastic and wire, shot through with ribbons of silicone and glass splinters. He'd only lost a week's work, having backed most of it up on his computer at UCL, where, for all I know, it still is, waiting patiently on some long-buried hard drive. (His department gave him sick leave, which I encouraged him to extend: I hadn't told Beth where my new job was, but she knew where Kit spent his days. When I fell apart, Kit's sick leave turned into indefinite, but unpaid, compassionate leave. I encouraged him to stay away; theoretically, he could still go back and complete his doctorate, but I have learned to stop reminding him of that.)

Both our mobile phones and our landline and answering machine had been destroyed, the latter a black pool of cooled lava on the table. We had already replaced them with new numbers. Almost all our books would have to go. The burning around the photo of my mum was so bad that I couldn't even identify which bit of charred mess had once been the frame.

Kit's coat had been hanging on the back of the bedroom door. All that remained were metal buttons, the skeleton of a Swiss Army knife and some blackened coins from his pocket.

He stood in front of where his eclipse map had been. My hand flew to my mouth. 'Kit! All your memorabilia!'

'That's what I thought,' he said. 'But actually most of it's in storage at Mum's. There's only really the T-shirts here and they might be ok.'

Amazingly, he was right. The bedroom was smoke-damaged but not fire-damaged. I opened the wardrobe and sniffed the nearest dress. It looked passable but the smell made me retch. 'We might be able to air our stuff, or wash it out,' I said. In silence, we bagged up the clothes and the contents of a bookcase that had escaped more or less unscathed, blackened spines wiping clean to reveal the titles beneath and the pages only edged with soot.

Everything else we left to be cleared by the landlord.

We bumped our bags down the stairs. Outside, the kebab man was sitting on an upturned crate, smoking a cigarette.

'It's a bad business.' He looked down at Kit's bandaged palm. 'Where you off to now? You staying local? Your mate was asking.'

'Mate?' My voice shook without my permission.

'That girl with black hair, she was here the other day. She was in a right state.'

'I'll bet she was,' I said under my breath.

'If she comes back,' said Kit evenly, 'tell her we're going travelling. Backpacking. Taking a gap year.'

'Yeah.' The kebab man nodded. 'Do you good, get a break from it all.'

On the way back to Adele's, we drove past Lambeth register office. We stopped at a red light. A bride and groom in shiny, round middle age laughed on the steps in a hail of rice.

'Let's do it now. What are we waiting for?' Kit's voice was musical with the thrill of uncharacteristic spontaneity.

'You mean it?' It was the first genuine smile I'd given in days.

'I'd fly you to Las Vegas tonight if we could afford it. But let's fill in the forms. Whatever step one is. I want you to be my wife. This has only made me realise how much it matters to me.' Determination had momentarily wiped his brow clean of the deep lines that had settled since the fire. 'Look; about the only good thing about the police not being able to charge Beth is that we're not tied to her by another court case. We can make a clean break. We get married. We start again. Change our names. Go and live somewhere else, maybe up where Mac and Ling are.'

The lights turned green without my noticing; I only broke our kiss to pull away when the driver behind blared his horn. I'd made my decision before I was out of first gear.

'Let's go for it,' I said. 'I can't think of a better way to disappear.'

The traditional red telephone box had the traditional stench of urine, and I had to breathe through my mouth. Outside, cars nosed along the still-unfamiliar neighbourhood of Green Lanes, past the Turkish bakery that never seemed to close and the forlorn jewellers that never seemed to open.

That evening I had finally, and with a breaking heart, come to terms with the cure for my insomnia. In the week since our move to Harringay, I hadn't slept for more than three hours. Exhaustion, if it continued at this level, was bound to loosen my tongue in a way beyond my control. I cradled the greasy receiver between my neck and chin, a five-pound Phonecard poised at the mouth of the slot. I had been in this position for five suffocating minutes, DS Carol Kent's business card balanced on my other palm. The card was ragged at the edges and grass-stained from Lizard Point. The

Jamie Balcombe Is Innocent campaigners would doubtless have wanted me to contact them directly but Kent's wrath seemed preferable to their glee.

I had known on hearing the verdict that Jamie's conviction was unsafe, but I had still believed it to be just. Now I knew that Beth, too, had lied to the police, and presumably again, under oath, about how she'd travelled to the festival, and while I could see her reasoning, it didn't chime with what I'd seen of her after the assault. In our flat, she had revealed herself to be first voyeuristic, then petulant, inconsistent and finally violent. None of these things individually undermined Jamie's guilty verdict, but collectively, they changed everything for me. My white lie was blackened, streaked with soot. Did I still believe Jamie was guilty? Yes . . . yes. Yes, most of the time. Did I still *know* he was guilty? No. The conviction was unsafe.

I knew what I stood to lose by making the call. The physical consequences hung over me. Perjury. Possibly perversion of the course of justice and contempt of court. It would be prison for me if I came clean, but this was as nothing compared to the personal consequences. I would be waving away my father's pride in me, Ling's respect; and almost certainly my relationship with Kit. The career I so desperately wanted to build was at stake too. I couldn't see many charities trusting a convicted perjurer with their reputations. I punched in the area code, every number a step away from the only life I'd ever wanted. And yet confession was what I would have expected, demanded, from anyone else in the same position. It was what Kit would have expected from me.

A siren shook the windows as a police car in chase forced vehicles to part where they could. As it inched past me, the long arm of the law got me in a chokehold. The muscles in my neck started to cramp and the walls of my throat closed in.

I couldn't do it. I saw, with horrible clarity, that I wanted my life and my reputation to stay the way they were. I never knew, till that moment, how big my ego really was. I consulted my reflection in a pane of dirty glass and felt a sudden, vertiginous loss of self. It seemed that I would rather take a chance, however small, on an innocent man serving time in prison than take responsibility for lying in court.

Staring into my own dark heart like that, is it any wonder I went mad?

At home, Kit was valiantly making spaghetti on our crappy little two-ring cooker, concern peeping out from behind his smile. He'll never look at me like that again if I tell him about the trial, I realised. I can never make that call. A strange feeling rode in on the wake of that thought; the hairs on my arms standing up in a wave, as though disturbed by a breeze, even though the air was thick with steam.

'This is good,' I said over supper. The pasta was slightly over-cooked, the way we both liked it.

'Thanks,' said Kit distractedly. His eyes didn't seem able to meet mine, but kept dropping to somewhere near my plate.

'What?' I set my fork down.

'It's just, could you stop scratching? It's really annoying.'

I followed his gaze to my forearms and was shocked to see them latticed with red lines. 'I didn't even know I was doing it,' I said, but I was suddenly aware of a low-level tickle on my skin. It felt as if I was walking through a wood in which webs trailed from every branch of every tree.

'Have we bought a different kind of washing powder or something? Maybe you're allergic to an ingredient or something.' He jumped up to check the cupboard under the sink. 'No, it's the same one.'

Now it felt as though the branches themselves were scraping at my skin. The scratch marks were rising into welts.

'What if it's some kind of delayed smoke inhalation?' I asked. 'Or nerve damage?' I didn't feel right on the inside either; there was a whirring in my chest, like something was trying to drill its way out.

With his good hand, Kit tilted my chin to look at my neck, then lifted my top to examine my belly and back. 'It must be a local-ised reaction to something. It's only on your arms.'

My skin was on fire all night. When I at last dropped off, I dreamed of my mother covering insect bites with calamine lotion, and woke with tears in my eyes and blood under my fingernails. The alarm clock display jumped from 8.20 to 8.21.

'Why didn't you wake me up?' I said, bursting into the sitting room where Kit was at his laptop, the modem lights glowing beside him.

'You're not going to work, I've called in sick,' he said. 'I've got you an emergency appointment at the GPs.'

'For a bit of *itching*?' I felt wired and edgy, like I'd drunk a whole pot of coffee.

He took my face in his hands. 'Whatever this is, we'll get through it together, ok?'

'You *do* think it's because of the fire?' I started to shake.

'You know I'd look after you whatever, don't you?' he said.

I only found out later that Kit had been awake all night looking up neuropathic itching on the internet, and coming up with a shortlist of degenerating conditions. He was as relieved as I was surprised by the GP's brisk diagnosis.

'You're having a panic attack, dear,' she said. 'Hardly surprising given what you've both been through.'

'No,' I said. 'This is *physical*.'

'It's psychosomatic,' corrected the doctor. 'The human mind's a tricky old sod. I'm going to prescribe a steroid cream to calm your skin and some Diazepam to give you a bit of breathing space, and I'm going to refer you to a counsellor so you can nip this in the bud before it gets out of control. It's a seven-week wait for counselling on the NHS. Can you pay to fast-track?'

All I could think of was the day's missed work, and Kit's hand, healing more slowly than the doctors had predicted.

'Yes,' said Kit instantly.

I beat the urge to scratch until we were on the pavement outside, when the itch was almost overpowering. I attacked already broken skin with my fingernails. Kit circled my forearms with his one good hand and held them fast when I tried to pull away. Beth's voice came to me. *You don't realise how much stronger than us they are.*

I couldn't get away from her. My thoughts folded in on themselves. Just because she's mad doesn't mean he *didn't* do it.

Would I ever be sure?

The counsellor they sent me to was good at overcoming the physical symptoms of anxiety; mindfulness techniques and exercises that could train my soma to outwit the psyche and get the prickling and shivering under control. But I could never tell her what I alone knew to be the root cause of the problem. The counsellor wasn't stupid: she knew I was holding back. Sometimes I thought about making up some childhood trauma to explain things away. In my darkest session, I thought about attributing it to grief at the loss of my mother.

Kit got me through it. He had saved my life twice: once in the fire and again, every day, every night, when he would get up with me and play cards and watch back-to-back *Seinfeld* DVDs in the small hours, brushing my hair, stroking it, plaiting it, while I fought the urge to scratch at my arms. I was so caught up in my own secret, self-imposed hell that it's only now I appreciate how much he sacrificed to look after me. It was out of the question that I continue to work, and no matter how much teaching work he took on there was no way he could support us both on the bursary for his doctorate. As soon as his consultant gave his scarred hand the all-clear, he took a part-time, temporary job as a technical assistant in a high-street opticians, snapping cheap lenses into spectacle frames where once he had stared through high-grade lenses at stars without name. He never once complained that it was beneath him.

Four weeks after their flat burned down, Laura Langrishe and Kit McCall, formerly of Clapham Common, SW4, exchanged wedding vows at Lambeth Register Office; Christopher and Laura Smith left the building and returned to their new rented home, a one-bedroom flat on Wilbraham Road, N8. There was no wedding breakfast, no carrying the bride over the threshold, but consummation was sweet and tender in the newly marital bed. The honeymoon was spent with the bride in twice-weekly psychotherapy, the groom working double shifts at his lab technician's job to pay for her sessions on top of the rent.

We were still only twenty-two and the quiet intensity of those weeks was too much for a couple of kids. While other couples our age were umming and aahing about commitment, we were so utterly enmeshed that we had the opposite problem. What is commitment to a relationship that is only darkness? Even sex, our one previous respite, lost its playfulness. We were all need and no want. When would the fun start again?

We had been living in Wilbraham Road for five months when Kit came home with a long thin tube under his arm.

'What's that?'

'Watch me.' He unfurled a map of the world and Blu-tack'd it to the blank wall over the fireplace. He'd bought – or more likely got from Adele – some red embroidery thread. Finally, he scissored off a length with a flourish and pinned a red thread across central and southern Africa.

'Zambia, January,' he said. 'There's a little festival. Just a few thousand people. Not Ling or Mac. Just us. We can keep to ourselves.' He smiled his old smile. 'There's virtually *zero* chance of precipitation.'

I knew that this was something we had to do if we were not only to survive but to thrive. That only by reclaiming the eclipse, and by standing under the shadow, would we again find our light.

40

LAURA
20 March 2015

'Why's everyone so obsessed with feeding me?'

Ling is as familiar with my kitchen as I am with hers; without looking, she picks a ladle from the drawer and slops chicken and sweetcorn soup into bowls.

'*Sit,*' she commands.

'That's easy for you to say.' Instead of a table in our kitchen we have a booth tucked into one corner, like something you'd find in an American diner, leather benches running either side of a Formica surface that's so old there's a half-inch gap at the join where it's peeled away from the wall, full of gunk and crumbs. I'll tackle it the week before the babies come, when everyone says I'll have a sudden urge to start washing windows and straightening cushions. I sit down and slide along the bench, back against the wall. I can see the whole kitchen from here. 'I'm not getting back in here till these babies are on the outside,' I say, feeling the squeeze.

Ling sets the bowl in front of me and watches me not eat, arms folded.

'I *want* to want it,' I say.

She lowers her brow in concern. 'Isn't losing your appetite one of the early warning signs?'

'Oh God, is it? For what?'

'For anxiety. Why, what did you think I meant?'

'What if the reason I'm not hungry is because the babies aren't growing properly? Or worse?' Ling doesn't need me to articulate what *worse* means. She was there for all the times it didn't take.

'Sweetheart, you're a nervous wreck,' she says. I am; she doesn't know the half of it. 'I was the same in both my pregnancies. It's normal, I promise. Look, we're literally on the way to get a scan. If it was an emergency they couldn't see you sooner. They'll put your mind at rest, I promise – are you even listening to me?'

I look up from my phone.

'Sorry. Yes. Thanks,' I say, but I know I won't relax until I hear Kit's voice. It's two hours now since the eclipse, easily enough time for him to find somewhere with a signal. He reckons that once the babies are here we'll take them eclipse chasing with us; plans are afoot for a family road trip to the States in 2017. If I'm this anxious at the thought of him on his own, how will I cope with guarding two toddlers in the middle of a crowd? I'll be too paranoid to look up, even for a second. I don't want to be one of those hovering, obsessive parents you see on the street, even though *my* fears are justified.

I push my soup away virtually untouched. On the way out, a delivery man intercepts me with a Mothercare package for Ronni next door.

'What if it's an omen?' I say to Ling. She takes it from my hands and squeezes.

'Feels more like a pair of wellies to me. Come on, let's get you scanned.'

The last few rounds of IVF were paid for privately, but now that I'm pregnant it's back to the NHS and the North Middlesex Hospital on the North Circular Road. How anyone's supposed to recover from anything next to the most polluted road in London I don't know. As usual, it smells of on-the-turn raw chicken and hand sanitiser. By the time we arrive, the babies are wrestling inside me. There is still no word from Kit and it is with reluctance that I obey the signs and switch my phone off.

My consultant, Mr Kendall, is a specialist in multiple pregnancies. He's been with me from the beginning and he'll deliver my babies. (Everyone, even Ling, assumes that I feel robbed of a natural birth. The truth is, it's one less thing out of my control, and I'm glad this decision was made for me.) Mr Kendall's flawless hands – he must have them professionally manicured – inspire confidence.

'No Christopher today?' he asks, nails shining as he taps numbers on a keyboard.

'He's gone off to watch the solar eclipse in the Faroes,' I say. 'Last big trip before the babies come.'

'Oh, yes,' he said. 'Always wanted to see one of those. We took the kids down to Cornwall in '99. Probably before your time. Total wash-out, anyway.'

I smile and lie down, preparing myself for the slick of cold gel. I'm an old hand at this by now.

'And how about you?' says Mr Kendall. 'When are you going to slow down?'

The truth is, I'll be working right up to the wire. We need the money.

'I'll stop soon,' I say.

Mr Kendall keeps the screen turned away from me as he slides the probe around the protruding nub of my navel and analyses the measurements. Kit would like this; this is the kind of quantifiable progress he can relate to. I must get a print-out of the numbers as well as the all-important picture.

'They're both growing fine,' says Mr Kendall. 'Your placentas are in the right place and so are the umbilici. You're *sure* you don't want to know whether they're pink or blue?'

I turn my head away. I decided to keep their sexes a surprise from the word go, to show Kit I am, in fact, capable of being laid-back. Although Kit has never called me a control freak in so many words, he's begged me to loosen up, slow down, take it easy and, even once, to our mutual horror, *chillax*, more times than I can count.

'Oh, well this one's definitely a . . .'

'*No!*'

Mr Kendall and Ling both seem taken aback by the violence of my reaction. I aim for a joke.

'Actually, we're going to raise them gender-neutral. Orange clothes. As an experiment.' It lands, but only just. It's a relief to get out of there.

Ling gives me a lift home.

'You sure you're ok on your own?' she says, even though I know she'll be working late to make up for accompanying me.

'I'll be fine,' I say, slamming the car door.

'EAT SOMETHING!' she calls, before driving off.

Back home, from the squidgy comfort of my sofa, I send a picture of the scan to Kit; he FaceTimes me by return. He's in a bar or something, dark shadows in the background, the rim of a beer glass just in shot on the table in front of him. The signal is patchy; his newly smooth face keeps turning into squares and the black and white pattern on his Faroese jumper jumps in and out of binary code. Even chopped into pixels I can see the drag of remorse on his features. The hard cold chip of my anger starts to thaw. I don't want us to argue again. For once, I'm the first one to end the sulk.

'Aren't they beautiful?' The warmth in my voice is the real thing.

'Everything's as it should be?' he asks. 'They're growing at the right rate? No horns or tails?'

'They're perfect.' He twitches a little smile. Only now do I remember why we're having this conversation on the phone and not in person. 'Sorry, baby, tell me about the eclipse. Were you completely clouded out?'

'Worst one yet,' he says dolefully. 'What about in London?'

'Shite.'

'That's a shame,' he says. Kit's voice is completely devoid of expression and my early warning system stirs into action. I search his face for clues but it's dark and the picture quality is poor.

'Something's wrong,' I say. 'Something's happened because of that sodding video.'

'It wasn't the video,' he says.

So something *is* wrong. I feel my skin start to prickle. 'What wasn't the video? What's going on?'

The signal falters again; his words click and screech like dolphin song.

'Look, don't worry. I'll be back on the boat soon, and then we can both relax.'

I scratch the arm that's holding the phone. 'Why can't you relax on dry land? Kit?' I can hear my voice going shrill and interrogative. I know it's like poking a snail's antenna; he'll only retreat into himself, but I can't help it. 'Have you seen her?'

'No.'

'Don't go all monosyllabic on me.'

'Stop over-analysing things,' he snaps back. 'You know me, I always get a bit of the old black dog when I've been clouded out.'

'Swear on our babies that nothing's happened.'

The connection cuts out for a second, so I can't tell if the pause before 'I swear' is his hesitation, my imagination or some kind of satellite delay, and then he's gone.

The magnitude of what I've done hits me. I've made him swear on our unborn children. Kit isn't superstitious and would say anything to talk me down from a panic attack. I have tempted fate. I put my hand on my belly and wait for the babies to kick, but nothing happens.

41

LAURA
20 June 2001

'Seven thousand totality freaks, all gathered together,' said a girl with a ring in each nostril and purple dreads twisted into upside-down cones on her head. 'It's gonna be fucking *filthy*.' She had a point; the location, deep in the Zambian bush, seemed to have filtered out both the casual ravers and the serious astronomers, many of whom disliked sound pollution almost as much as light pollution. Kit and I were five hours into a six-hour journey from Livingstone Airport to the festival site, in an unventilated bus with poor suspension and fifty hippies who used only natural deodorant. Scrubbing-brush trees sprang green against apricot soil. Cattle walked between speeding, laden trucks without flinching. Roadside greengrocers displayed rainbow wares. Old painted advertisement hoardings and towns made of corrugated iron came and went in the turn of a wheel. When we stopped for lunch in a Fanta-branded café, dozens of children emerged as if from nowhere, and all of them wanted to touch my hair, shrieking with delight as tiny fingers tugged and twirled it into knots.

Civilisation seemed a long way away by the time we crested the hill to see the makeshift township. The festival site had better infrastructure than some of the villages we'd passed en route, with a little bush supermarket, a row of showers and African-style squat-down toilets that were cleaner than any you'd find in a

261

British shopping centre. Cafés sold drugs as openly and cheaply as beer.

'It feels right,' I said to Kit. He simply tilted his face to the sky and grinned. There would be no checking the forecast for him here, no worrying about weather conditions. An African winter is predictable. That day, the sky was the kind of warm bright blue that makes you doubt the colour could ever be associated with coldness, and its clarity made us doubt the very existence of clouds. On a fully rigged stage, a reggae band chugged their way through a series of Bob Marley covers. We walked past a tent where a marbled graphic was being projected on to one wall. Kit picked up a little coconut mask of a screaming African face and held it over his own eyes. 'This looks exactly like Mac did when he was sectioned.' It was the first time he'd ever been able to joke about Mac's recovery.

There was no dusk; one minute the sun turned the colour of orange squash and the next it sank without warning behind the horizon. We ran through the new night to pile on the layers and lie at the mouth of our tent, backs on the ground. At 4,000 feet above sea level, with minimal light pollution and only a pink moon glowing, the African sky was not so much speckled with stars as streaked with them; the effect was meteorological rather than celestial, the Milky Way a stormcloud that threatened to rain glitter. There was an atavistic satisfaction in gazing at the heavens.

'We're meant to look up,' I said. He nodded into my shoulder, then rolled into me. The shallow convex of his belly tucked into the small of my back, matching the rise and fall of my diaphragm. Even our breathing was in synchronicity that night.

Perhaps it was the effort involved in getting here or my need for atonement but Zambia felt more pilgrimage than holiday. We eschewed the stages and bars for the shade of trees. That must be why it took her so long to see us.

When it was time for the eclipse, the organisers pulled the plug on the sound systems. 'First contact,' whispered Kit, as the moon took its first baby bite out of the sun, to soaring cheers. Here at last was the reverence I had so hoped for in Cornwall. Not a silence, there were too many of us for that, but every whoop and whisper was respectful of and somehow *devoted* to the phenomenon. The crowd tilted their faces towards the shrinking crescent sun in heliotrope unison. An hour followed in which it seemed to me I barely blinked.

'Don't you want to get a picture?' I asked, gesturing to the camera dangling by his side. He waved it away.

'Nah,' he said, to my astonishment. 'Not this time. Let me just *live* it for once.'

Then the winds came, that eerie nowhere breeze I'd experienced in Cornwall, only here it was warm and gritty with dust. As if at the wind's signal, time, crawling for the last hour, sped up and dusk pushed in from the east.

'*This* is what two thousand miles an hour looks like,' said Kit, as the darkness charged us. The landscape changed as dramatically as the sky, our shadows suddenly shrinking as though we were melting into the ground. I started to shake, and reached for Kit's hand. He lifted his arm and twirled me around, ballerina-style, so that I could see the twilight on every horizon.

A ring of white light surrounded the sun, one pure diamond flare teetering at the tip, and then the switch of totality was thrown. *Oh my God, dio mio, mein Gott, wow*, screamed the crowd. The moon was a black disc covering the sun and streamers of plasma flared out, like a gas ring being ignited. 'Is it safe?' I asked Kit. I meant can I take my glasses off but the question felt bigger, too. Is it safe to be alive, on this spinning rock? Is it safe to be this small? Are we going to be all right?

He removed my glasses in answer and I looked with naked eyes at the coal-black ball in the sky. I knew all the theory, I knew I was looking at vast promontories of hydrogen gas, but as I stood there I could think only in terms of gods and magic. The corona danced, a living golden flare twice as big as the sun itself. A star is not an angel but a monster. It was so huge that it made everything that had happened to us, everything we had done, seem tiny. Regret, guilt and fear melted away.

'I'm *healed*,' I said, and it didn't feel trite in this context. You can say anything under the shadow. Kit's cheek was wet on my shoulder and I matched him tear for tear. We weren't the only ones crying; there was soft sobbing nearby, and in the distance, someone howled like a wolf. We stayed like that for four and a half minutes. As if programmed by some internal clock, Kit slid my goggles back down like a visor; seconds later, there was a blaze of yellow light and the shadows melted away to the east. It was over. The tears I wiped away now were happy.

'When's the next one?' I said.

The following day, buses that had brought us here arrived to take us back, only now there were seven thousand people all trying to leave one destination at one time and chaos ensued; some were bound for Lusaka Airport, others going further south to Livingstone. Some poor sods from Japan faced two days on the bus to Johannesburg. Kit and I queued on one side of the road for the Livingstone bus, while the opposite queue for the Lusaka shuttle snaked for half a mile. After a two-hour wait it was finally our turn to lug our backpacks on to the roof-rack and take our places on the sticky plastic seats. I wiped the window down but the dust was on the outside.

'I don't want to go home,' I said to Kit. 'I just want to do it all again.' I said this knowing full well that the next one was in the Antarctic circle.

'I can't see many of this lot making the South Pole.' Kit surveyed the gurning casualties around us. Our bus's engine was still idling even though all the seats were taken. 'They couldn't afford it. No humans have ever witnessed an eclipse from the pole before. It's a massive expedition, it'll cost thousands. I can't see *us* making it.'

I don't know what made me look back. Some avenging spirit of justice, punishing me for daring to drop my guard and be happy.

But I did look back, craning over my shoulder at the other bus. She was framed by a grubby window, staring at me, into me.

'*Beth*.' It came out of me in a growl. There was a moment of absolute stillness and horror. Kit froze, then slowly followed my gaze. The three of us were locked like that for a heartbeat. Then, like a spider after a fly, she started to move.

'Can you get going?' said Kit to the driver. On the other bus, Beth clambered over her neighbour and began to move down the aisle, clearly obstructed by bags or bodies. The driver was the embodiment of the word 'laconic'; he raised an eyebrow at Kit and sucked his huge yellow teeth.

'Just fucking drive!' said Kit. I'd never seen him look so frightened, not even in the fire.

'Please,' I begged the driver, somehow mustering a calm smile. 'We really need to be somewhere.' I fished in my pocket and found a handful of *kwacha*; Kit threw the notes into the driver's lap. Finally the wheels spun, churning the thin soil, so that when Beth made it into the stony road, she stumbled into a cloud of exhaust and swirling bush dust. A motorbike roared past her, so close I thought it would skin her knees. Undeterred, she tried to run alongside the bus, but the flip-flops on her feet slowed her down. 'Laura!' she said, but there was no threat in her voice. She looked desperate rather than angry, aggrieved rather than frightening.

'Don't slow down, don't stop!' said Kit to the driver. 'Put your fucking foot down!' It was the first time he had shown his fear in front of me. Only in the stripping of his strength did I understand what it had cost him to carry me. The bus surged forward. I turned around, stood up in my seat, and saw Beth collapse to her knees in the street, her skirt puddling around her legs so she looked like a landed mermaid, suffocating in a swirling red cloud. We rounded a bend and she was gone from my sight.

Everyone on the bus was staring at us. Kit sank low in his seat, glowing with embarrassment. The silence ensued until the driver began to play with his radio. 'Livin' la Vida Loca' was playing and the driver sang tunelessly along.

'She came halfway across the planet to find us,' Kit said in a voice so low even I could barely hear it. 'All this effort we've gone to in changing our names, making a new life, it's all been for nothing. And we told her where we were going to be.'

The bus rolled past patchy fields of skinny cattle. I took hold of Kit's bad hand and ran my fingers over the scar. He pulled it away. 'She saw my map, she knows where to find us for ever.'

He didn't say it but the implication was clear; because *you* invited her into our house. I reached for his hand again, but it was clenched in a fist.

42

LAURA
15 November 2003

Jamie Balcombe was released after serving half his sentence. He had been a model prisoner for two and a half years. In the prison library, he taught fellow inmates to read and write. After the fire he wrote me one more letter, saying he had heard about the incident – I shuddered at the knowledge that he was keeping tabs on us – and hoped that my own experience of trauma had increased my empathy, and the likelihood of my retracting my statement. I re-sealed the envelope and sent it back to the prison, saying that there was no forwarding address, and no more letters came. I don't know if the authorities opened it; if they did, it didn't count against him when it came to probation.

Jamie's freedom brought to me both fear and release.

Fear that he would somehow get through to Kit. I was confident Kit would believe my lie over a convicted rapist's truth, but still it remained a lie, and not one I could bear to repeat to the person whose opinion mattered most. And release, because I had felt every day of his imprisonment like a score in my skin. At least now he wasn't in prison; it was only – only! – his good name that remained tarnished.

We didn't see another total eclipse for five years after Zambia. We were – well, I was – too scared of Beth to go to South Africa in 2002. Antarctica in November 2003 might as well have been a flight to the moon as far as we were financially concerned, coming the year after we'd bought the house on Wilbraham Road, lying on the then-ubiquitous self-certification mortgage application to make it happen.

We still chased the shadow, but gone was the excitement of the big alternative parties; instead, we slid under the radar. In 2006, when the path of totality was thickest across Libya, and the third 'festival of a lifetime' was held in Turkey, we saw the eclipse at the opposite end, in Brazil.

The night before our flight, Kit caught me sneaking hydrocortisone cream and a whole blister pack of Diazepam into a side pocket of our bag.

'We don't *have* to do it,' he said, searching my face.

'It's all booked. It's all paid for!'

It *wasn't* actually paid for, which made it worse; the whole trip had been booked on already strained credit cards. Kit smiled at me with such obvious effort I thought his cheeks might creak.

'It doesn't matter,' he said. 'If it's going to make you ill again, it's not worth it.' I was touched by how hard Kit tried to make me believe him, but also I knew him better than that. Foreign travel and mind-blowing primal experiences would have appealed to *any* man more than a nervous wife with scabs on her arms, let alone Kit, whose compulsion to chase eclipses had been in him long before he met me. In my lowest moods, I worried it ran deeper. These dates were in his DNA; they were his last connection to his father. He would hold it against me for ever if I was the reason he had to stop. It was my fault, my lie, my bad judgement that had let Beth in. It was my responsibility to deal with it.

'Of course we're going,' I said. 'We can't let her win.'

'Thank you,' he said. 'I know what it took you to say that.'

He knew part of it, I suppose.

We stayed in a budget hotel and watched a clean, perfect eclipse together, sitting on the bonnet of our hired car, halfway up a hillside where we were the only English-speakers for miles around. Four minutes, seven seconds of totality: the moon covered the sun, ripping a huge bullet-hole in a gigantic canvas. After that, we could believe that Beth had given up on us. Of course we would eventually find out that she had been in Turkey. In 2006 YouTube was in its relative infancy and it was another couple of years before a German clubber posted *the video* online.

Our subsequent trips were poisoned by my anxiety to various degrees.

Kit tried to make it easy for me when we went to China in July 2009. We booked into an anonymous hotel next to a motorway, the last place she would look. Even so, I barely slept the week before we travelled. Knowing I had to tolerate the discomfort didn't stop me panicking in the airport and I took so much Valium on the outbound flight that Kit had to carry me off the plane. It was the uneventful eclipse – if such a thing is possible – that he had promised. Most of the trip was spent watching Kit fiddle with photographic equipment next to a motorway.

I would have loved to go to the huge festival they held on Easter Island for the 2010 total eclipse, but Kit said it wasn't worth my paranoia, and so we saw it halfway up the Andes in Patagonia. On the mountainside the snow was so dry it felt more like sand than water. Kit got some beautiful shots of the shadow cone spreading out across the snow. I could have sworn I didn't take my eyes off the sky throughout totality, but our guide took a picture of me, looking over my shoulder, when all around me were gazing up in rapture.

I ought to have relaxed in Cairns in 2012; tens of thousands of people watched from the miles of palm beaches on Australia's Gold Coast, but by then smartphones were everywhere and, terrified that someone would broadcast us as they had Beth, I wore a hat so huge I could barely see the sun. Now, when I look at the globe and envision the path of totality for that eclipse, it seems to me half of Australia was covered by the shadows, that the chances of her finding us on such a huge land mass were tiny. Not like now, when the shadow falls mostly over water, and everyone is squeezed on to those dark northern islands, so tiny that there is nowhere truly to hide.

43

LAURA
20 March 2015

Now the sun decides to show itself. It is early evening and it shoots through streaky pink and purple rashers, the kind of clouds that see cameraphones all over London pointed at the sky.

I wonder if the weather has cleared in the Faroes. In an hour Kit will be at sea. There can be no passengers on the *Princess Celeste*'s return leg who weren't on the outbound journey. It's a chartered package tour, not a scheduled ferry. Perhaps she was there, or perhaps she was somewhere – on Svalbard, maybe, or on another island. Or perhaps she was in Tórshavn itself; maybe she watched the non-event on the other side to Kit of some crag or rock, eyes not on the sky but with an old picture of us in one hand and a screen-grab from that vlog in another. But if she was, she didn't find him.

Tomorrow lunchtime, he will be home, with a load more souvenir tat. He'll have his camera plugged into the laptop, uploading his photos, before he's even got his coat off. With every passing minute, my fears seem unfounded and my outburst the last time we spoke to each other looks more and more like paranoia, me reading darkness in what was probably just a bad mood. Even the babies inside me seem calmer.

It can't be more than an hour since I last ate but I'm ravenous all over again. I chew the last hunk of Mac's sourdough and wash it down with the only drink that's left, some kind of livid berry concoction; a film of green powder floats on its surface like algae. It probably is algae, organically sourced, local, Fairtrade algae. God knows what he charges paying customers for one of these. My children had better come out bursting with health after all the vitamins I'm taking.

The BBC are doing a special programme about tonight's eclipse and even though I'm recording it for Kit, I watch it anyway, keen to picture, now that it's not live, the place he's been for the last few days. I've just about lowered myself on to the sofa when the doorbell goes. I heave my bulk out of the cushions and pick up Ronni's Mothercare parcel, ready to hand it over. I feel it, like a kid would a Christmas present. Ling was right. It *is* wellies. Maybe I'll invite Ronni in for a coffee, if she isn't busy putting the kids to bed.

I forget that the chain's on the door. I pull it two inches wide to see Beth Taylor standing on my doorstep.

The cup falls from my slackened grip and the purple juice explodes across the floor tiles and up the walls, making the floor slippery beneath me, the tart berry smell at odds with the metallic fear on my tongue. Instinctively my hand goes over my belly and I stagger backwards, catching the banister to stop me falling as my foot skids on the mess beneath me. I find myself sitting on the bottom step, hands pathetically trying to cover my bump. Whatever Beth was going to say, the shock of my fall and, judging by the way she's staring at my belly, of my pregnancy takes her breath away for a few seconds, then she says, 'Shit, Laura. Are you ok?'

'What do you want?'

She puts her hand through the gap between the door and the jamb, a stretching motion that could either be a pathetic offer to help me up or a hand curling around to undo the chain. It will only fit as far as the wrist.

'Laura,' she says.

'No!' I say, and she draws her hand quickly back.

I look at the console table but the landline handset isn't in its cradle; too late, I remember leaving it upstairs when I spoke to Dad this afternoon. My mobile is somewhere in the sitting room, and in the time it would take me to go and get it, who knows what she's capable of? Unhooking the chain? Kicking the door down? Has she got a weapon?

'Please, Laura,' she says, pressing her face up close to the gap between the door and the jamb, so all I can see is her nose, her mouth and the inside corners of her eyes. 'I really need you to let me in.'

Third Contact

44

KIT
20 March 2015

'Svalbard,' says Richard, waving his tablet under my nose. 'We should've gone to bloody Svalbard.' We've found a bar in the harbour, with free Wi-Fi, a view of the red house and Sólarbjór on tap; my last drink for a while, I think, and decide then and there not to touch any more booze until the babies are safely born. Richard has decided that our last few minutes on the Faroes should be spent bent over a screen, wallowing in what could have been. 'Or even some of the other Faroe Islands – this was taken just twenty miles north of here.' He zooms close on a beautiful image of a crescent sun in a clear sky. 'And look! There was a cruise ship that couldn't dock in Tórshavn and they all saw it from the middle of the sea! Why didn't the *Celeste* do that?'

'Why don't you sue them for bad weather?' I suggest. You do actually hear of that happening; tourists, not serious chasers, haranguing their tour operators for their money back if they're clouded out. Richard takes my point.

'Speaking of the *Celeste*, we should get a move on. Gangplank's up in half an hour. Have you got your stuff?'

'Indeed I do,' I say, patting down the paper bag I've just acquired. My 'stuff' is two miniature T-shirts with a picture of the corona on them that I'm pretty sure came from the 2006 eclipse, I can tell by the distinctive horizontal flare of the hydrogen promontories.

The T-shirts look tiny to me, too small for anything but a doll or a teddy bear, but the label inside says 12–18 months. I am struck again by how imminent fatherhood is, and how little I know about babies. I thought I was quite a hands-on uncle with Juno and Piper but Ling and Laura tell me that I used to let their heads wobble all over the place, and that I always handed them back when they started crying.

I stand up, feeling the camera strap cut into my neck and, draining the last of my Faroese beer, strike out after Richard for the ship. I'm two drinks down, and it's partly the alcohol, but a kind of euphoria is moving in to replace the paranoia, the *madness* that infected me up on the mountain. All Laura's fears have come to nothing and in twenty paces I'll be on the boat and on the way home to her.

'Kit!'

The woman's voice is right behind me, so close that she didn't need to shout. Richard rolls his eyes, bored now of my celebrity, but my heart hammers. The shock of hearing my real name, the one that Laura calls me, in this context, stops time then stretches it. A hand is on my arm before I register something important: although it's a female voice, the accent is wrong. I turn to face a woman about my age with dark hair, a washed-out prettiness, smiling at me with delighted recognition. I don't have a clue who she is, although this doesn't seem to faze her.

'It's Krista!' she says. 'Krista Miller!'

My mind scrambles vainly for a foothold.

'I'm so sorry, Krista,' I say. I hope my smile is charming. 'You'll have to help me.'

'Ah, it's fine,' she says, totally unfazed. 'I've changed a lot – although you haven't. Aruba? Aruba '98? Baby Beach?'

A memory forms, slow as a Polaroid. The last eclipse I saw with my dad. Missing a week at university – it was the year before I met Laura – to fly out of the UK in February and lie on white sand that rippled like silk and burned like hot metal. A *beautiful* eclipse, Venus and Jupiter shining studs in the sky. All the Brits and Americans congregating at the same beachside bar every night of the trip and among their number an American student in – the image coalesces quicker now – gypsy top and Bermuda shorts, train-track braces. I remember posing for a group photograph that someone had taken with one of those disposable cameras that were all the rage back then. Memories flood back of promising to write, to get a copy of the photo sent, of street addresses rather than emails being exchanged. Lost in the past, I become aware that Krista's staring expectantly at me.

'Of course I remember,' I say, and reach down to return the hug she offers.

'I can't believe my luck, catching you like this,' she says. 'I was showing your photo around but I can't believe it's you. We'd all given you up for dead!' She laughs in delight. 'You know a whole bunch of us kept in touch after Aruba? I did write you . . .' No wonder she couldn't find me. Everything would have been sent and received via my childhood home. My parents moved so often that the post didn't always catch up, and now there is no longer a Kit McCall to trace. 'I tried to contact you through Facebook, or I thought it was you, I remembered you were wearing that same Chile '91 t-shirt and you got a huge . . .' she lowers her voice here, as though the drug squad have been waiting all this time to bust us, 'You got a *burn* from a *joint* on the front of it; you were devastated.'

'ShadyLady!' I say hearing the unintentional comedy in the words on their way out. 'You should've told me your name, said how we knew each other!'

Krista slams her palm into her forehead.

'No wonder you didn't write me back. That stupid name. My fault. I should have been more persistent.' She shrugs. 'We had a big get-together the other night. Are you free this evening? Some of the other guys are around.' Suddenly she registers Richard's presence. 'You too,' she says with a smile. Richard shifts unsubtly from foot to foot.

'My boat goes any minute now,' I say, gesturing to the huge vessel behind me.

'No way!' Krista wrinkles her nose. 'But let's keep in touch, we're planning the *mother* of all reunions for 2017,' she says. 'We live right in the path of totality. You should camp in our yard.'

'I'd love that,' I say, and realise I mean it.

'I married Bill, you know,' she says. She pulls a 'whodathunkit' face at me, and I mirror it back, thinking, Who the fuck is Bill? I have no idea. 'He's literally along any second now, you *have* to wait and say hi.'

I look at my watch; ten minutes till they close the doors and the boat is only a few steps away.

'Wow! It'll be good to see him again,' I say, still none the wiser as to Bill's identity.

'I'll see you in the cabin,' says Richard. 'Nice to meet you,' he says to Krista, and only now do I realise I haven't introduced them.

'I thought he was your brother for a few seconds,' she says. 'How *is* he? Still wild?'

'Not so much, these days,' I say. 'But he got a lot wilder before he got calm. He's good. Couple of kids, runs his own business in London.'

'And how about you?' she says, although I've already seen her glance down at my ring finger.

'Married my university girlfriend. Laura.' Saying her name casts a little hook in my heart, pulling me home. 'She's at home in London, pregnant with twins. It's been a long haul. Four rounds of IVF.' To my horror, my voice starts to crack. I don't know if it's the release of tension or the shock of a face from my youth, but there's something about this familiar stranger that makes me want to tell her everything. All of it, even the stuff I've never told Mac.

'Four years,' says Krista, while I work hard to control my face. 'That's hard, on both of you. But twins, though! Two little eclipse chasers to cart around the world. We've taken our two everywhere. You think it's disappointing when you're clouded out, that's nothing compared to the look on kids' faces. Oh look! Speak of the devils.' She waves manically – I'm remembering now that Krista did most things manically – as the fabled Bill shoulders his way through the crowd towards us, leading two little children, a boy and a girl. They're all wearing the same violet anoraks as Krista. It's a snapshot of my own future; any sense of style utterly surrendered to family life. They are the picture of belonging.

'Buddy!' he says, delightedly to me. I've never seen this man before in my life.

'Good to see you again!' I say. 'It's been too long.'

Bill looks skyward. 'Too bad we were clouded out.'

We compare notes – they took their chances in the harbour, didn't see anything either – and then swap email addresses, and hug goodbye with promises to write to each other soon.

It's with a sense of relief that I board the *Celeste*. If I can make myself this visible over the course of a few days, and Beth hasn't found me, then the danger *must* have passed. Beth didn't find us here, when I might as well have carried a flashing neon sign above my head for three days. She must have given up.

45

LAURA
20 March 2015

'Please don't shut the door,' says Beth. She leans in, blocking my view of the street. 'Seriously, are you ok after that fall? I didn't know you were pregnant.'

I wipe a smear of berry juice from my cheekbone; it's in my hair, too, I realise, slapping a wet purple strand over my shoulder.

'How did you find me?' Too late I realise that I should've said *us*; how did you find *us*? I mustn't let her know I'm at home on my own. No, it's too late for that; she must know Kit's away. It must be that fucking video, although I can't think how she made the leap from his location on board the *Princess Celeste* to Wilbraham Road. My heart is sprinting and my thoughts can't keep up.

'If you let me in I can tell you,' she says. I don't respond, not verbally and not physically. My mouth might as well be sewn up; my feet might as well be nailed to the floor. All I can do is stare.

Beth looks good, much the same apart from her hair. A shooting star at one temple sends a debris of silver through the black. The smile that dares me to drop my guard brings out faint crows' feet around her eyes, but otherwise her face is the same. The door shudders a little and I look down to see her toe – I'm half expecting those silver trainers but it's a clumpy boot, the kind I wore back in the day – edging itself into the crack. All that stands

282

between me and Beth is old glass and splinters. Action comes to me at last; I slam the door in her face with a force that shoves her toe out of the way.

'Ok,' she says, through the stained glass. There is no anger in her voice; only resignation. 'I was expecting this. But I do need to talk to you. I've got something to tell you and it's for your own good, I promise. Half an hour, let me say my piece, then I'll go. It's not exactly a conversation you want to have through the letter-box.' Her tone grows indignant. 'I'm not here for *fun*.'

'How did you get this address?' I ask her again, but still she doesn't answer. I'd forgotten how she shuts down questions she didn't like.

Horribly aware of the fragility of my front door, I inch into the sitting room. There, in relative darkness, I press nine on my mobile keypad. I've only called the emergency services once in my life and that was on the Lizard. I hit the number again. Can I really do this, now? After what I did in court, this number has felt off-limits; not a lifeline but the spark on a fuse that will blow everything apart. One investigation will lead to another and my perjury will come out. Kit will find out. He can't know that about me. The whole world will know; Jamie Balcombe's team will make sure of that. With so much at stake, such a controversial conviction, any judge is bound to make an example of me. What if they put me in prison? What if they take my babies away, or lock me up? The thought sends a slow cramp around my belly, as though my muscles are locking them tighter than ever inside me. I delete one nine, then the other. The letterbox rattles as Beth hooks it open with a fingertip.

'Lauraaaaa,' she sings.

The only person I want to talk to is Kit, but even through my panic I know that's a bad idea. All it will do is make him frantic with worry, and it's not like the captain can sail any faster. I want my dad, but he's miles away too and the thought of explaining it all to him is nearly as bad as Kit learning what I did in court. I call Ling but her phone's switched off. She does this when she's somewhere sensitive, like a police station or court. After this afternoon's scan, she probably thinks everything's all right. Three hours ago, everything *was* fine. In desperation, with no idea what I'm going to say, I call Mac. For all his faults, he is loyal and I know that he, more than anyone I know, will help me first and ask questions later. 'The person you are calling is on the phone,' says the droid voice. 'Please leave a message after the tone.'

'Mac, I need you to ring me back, now.' A sob knocks its way into the last word.

'Laura,' comes Beth's voice through the letterbox. 'Just hear me out.'

'The police are on their way,' I lie.

'Fine by me,' she says. I'm thrown by how calm she sounds. Is she bluffing too?

I give Mac ten more seconds to call me back, then fire off a text.

MAC PLEASE PLEASE PLEASE CALL OR COME OVER ASAP. XXX

I'm sweating all over. This is adrenaline, toxic to my babies' well-being and possibly to my survival. Fight-or-flight is a physical thing and it's my mind I need control of now. Crossing the hall tiles, I use mindful walking – heel, arch, toe, heel, arch, toe – to beat my panic attack.

Beth comes back to standing, pressing her hand against a royal blue rectangle of glass; I'm almost close enough to read her palm. I sit on the bottom step staring into my silent phone. I wonder about making a break for it, through the back door, and across Ronni's garden but I'm in no shape to vault the six-foot fences that separate the gardens round here, and anyway there's the river in one direction and in the other is Harringay Passage, the alley that comes out pretty much where Beth is standing. Her hand pulses a few times on the glass, like she's counting in her head to keep her temper. When she speaks again there's a new edge to her voice.

'Have you got a smartphone?'

It's glowing in my hand; she can probably see it through the glass. 'Yes,' I say in a small voice.

'Google Jamie Balcombe now, while I wait. I don't mean go to his website. Put his name into Google.' Her hand flexes one more time. 'Cross-reference it with court.'

Not retrial but court. The surprise disarms me into following her instructions. The phone buffers, the little pinwheel on my screen turning for an eternity. I can hear Beth breathing; I can almost feel the heat of her.

'You got it?' says Beth.

'Hang on.'

The pinwheel stops.

The results distil into headlines.

SIX MONTHS FOR DISGRACED CAMPAIGNER
NEW SHAME FOR JAMIE
MAGNATE CHARGED WITH ASSAULT
RAPE TYCOON ON BATTERY CHARGE
JAMIE: YOU HYPOCRITE

The letters seem to pick up their legs and march words across the page.

How the hell have I missed this? How did I take the website at face value? I've been burying my head in the sand since I got pregnant, that's the problem. This has nothing to do with Cornwall, nor even with Beth as far as I can see, so what can it have to do with me? Profound but futile relief wells up inside me; what I need now is to understand.

'Christ,' I say.

'Let me in?' She's almost begging.

I think about what it was like in my burning flat, how the flames grated me from the inside. I think about how Kit's scarred hand feels in mine even now. 'Beth, how can you even ask that? How could I *ever* let you into my home again?'

She clicks her tongue, then mutters something under her breath, of which I only catch the last two words: 'that again.' Her hand peels away, leaving a faint print of grease and sweat, and she bobs down from the window. 'I knew this would happen. Ok, I tell you what,' she says. There's the sound of rooting through a bag. 'That old pub at the end of the road, the Salisbury, I'll wait for you there. I've been here since the crack of dawn, I might as well stay another hour.'

'What do you mean, crack of dawn?'

Has she been waiting outside all day? Was she here for the eclipse? Was she watching *me* during the eclipse; did she follow me out to Duckett's Common? Was she here when the *girls* were here?

'I really can't tell you everything here,' she says. 'Look, I've got some stuff for you to read. I came prepared for this.'

There's the rustle of paper, then a white A4 envelope pokes through the letterbox, unsealed and stuffed to the point of stiffness. There's no name on the front.

'It's half seven now. Promise me you'll read this. I'll stay in that pub till half eight, you'll need a while to get your head around it. I'll tell you everything then. OK? I'll answer any questions you have. All right?'

'Ok,' I say, because what choice do I have? There is so much I need to know, about Beth and the past and what the hell all this Jamie business has to do with me now. Beth puts her hand back on the glass, soft as a caress. I wonder if she knows how little pressure it would take to break through it. 'Laura. I want to help you. Please don't act like I'm trying to do you harm.'

The puff of black hair disappears through the coloured glass, and she is gone. I'm left alone, in my house, which is covered in dark purple juice, as am I. It's up the walls, all over the floor tiles and in my hair, and the sweet berry smell is already starting to cloy. I ought to clear it up but the clock on my phone tells me that it's already 7.31 and there's a fat stack of paper stuffed into this envelope. With an inelegant thud, I sit back on the bottom step; the envelope trembles in my hand and it's an effort to slide the contents out smoothly.

The clock jumps to 7.32.

46

LAURA
20 March 2015

When my phone rings out and Mac's picture fills the screen, the surprise is electric, even though I begged him to call. My saviour of only a few minutes before is now an interruption I can't afford.

'Laura, what the fuck? Are you in hospital?' Mac sounds just like Kit when he's panicked, the drawl all gone, and I can only just hear him over a background of clattering and voices.

'I'm fine,' I say. 'Babies are fine . . .' There is a loud bang in the background and someone shouts *Bollocks!* 'What was that?'

'Burst pipe in the shop,' he says. 'Basement's knee-deep in water. I've been down here trying to sort it all out, that's why my phone was out of range. All the stock's damaged. The clear-up's going to take me through the night.' He pauses, finally catching up with the conversation. 'But if the babies are all right, what's up, why'd you ring?' His voice falls off a cliff. 'Shit, it's not Kit, is it?'

The irony makes me want to howl with wild laughter, that all the time I've been worried about Kit, he's been safe in his floating fortress.

'He's fine,' I say. 'I'm sorry, it was a false alarm,' I won't tell him it was anxiety because he will only fob me off with Ling and I need as many people as possible on standby for me.

'I'll keep my phone upstairs.' The drawl is back but there's the brittleness too now of forced patience. 'I should hear it if it rings. And I'll keep checking it.'

It's a relief to get him off the phone. It's 7.40. Beth will be in the Salisbury now, a glass of wine or coffee on the go.

The topmost page, a loose two-sided letter, must be the thing I heard her stuffing into the envelope. I flick quickly through the rest – photocopies and internet print-outs – before going back to the start.

Dear Laura

I'm writing this at my kitchen table before I come down to London. If you're reading this, it means that for whatever reason we haven't been able to have a proper conversation. Maybe I haven't found your house. Maybe I have found it but you don't want to talk. I'd wanted to go through all of this side-by-side, talking you through what it all means. But a letter through the door is better than nothing. What matters is that you get the information. It's been eating away at me for weeks.

I know that hearing from me out of the blue like this will shake you up. Trust me, dragging up our history is not exactly my idea of fun either. I thought the whole Jamie thing was dead and buried years ago, and I don't appreciate that all this time later it's coming back to bite us again. But there you go.

Please read everything carefully. There's quite a lot of it, but I've tried to get it in some kind of order that lets you know what's been happening.

Don't worry, I'm not interesting in rekindling our friendship. You've made it pretty clear you want to wash your hands of me and to be honest after a while I ran out of energy trying to find you, and persuade you otherwise. Anyway, whatever happened back then, I'll always be grateful for what you did for me, and that's why I'm getting in touch now.

I do think you should take the threat seriously.

With love,

Beth Xx

Her letter raises more questions than it answers. What's been eating away at her for weeks? What threat? *How did she find me?*

I turn over to the first of the collated pages. Next is a photocopy of a typed letter, written six months ago; the week I found out I was pregnant. The week Jamie's website changed.

> Antonia Balcombe
> c/o Imrie and Cunningham Chambers
> 198 Bedford Row
> London EC1

Elizabeth Taylor
c/o Evans and Bay Chartered Solicitors
1 Broad Street
Gedling
Nottingham NG15

3 October 2014

Dear Beth,
I hope you don't mind me calling you Beth. I hope that you receive this letter safely, and that you are amenable to reading it. I hope a great deal at the moment, it seems. I don't deserve your attention and I wouldn't blame you if you wanted to burn it, or even throw it in the bin, but I hope you read it. I hope.

You might have noticed that last week Jamie's website – that hateful, blasted website – was replaced by a notice that there had been a new development in the court case. This isn't true. How can it be, after fifteen years? What has happened is that, after years of misery and abuse, I have finally found the courage to leave Jamie, and am no longer prepared to put my name to his cause. But I have not withdrawn my support because we are divorcing; we are divorcing because I have withdrawn my support. Support I should never have given him in the first place. To say I owe you an apology is an insult.

I have been determined that this letter will not descend into excuse-making. But I do feel I need to explain.

Jamie is a bully, and I suppose I must be bully-able. He's clever; he's charming. It sounds silly writing it down but admitting to 'infidelity' with you was his masterstroke. I thought it was honesty; I didn't realise it was just the boldest of his lies. You must have been so strong, to stand up to him. When I think back to the trial, which happens often, I admire you so much and eventually your bravery played a big part in my own. I have a new vocabulary now; all that time living with what they call coercive control and even though I could have filled a book with examples of it, I didn't know it had a name. I thought it was just something <u>he</u> did to <u>me</u>.

There are levels of knowing things, Beth. And I think that even then I knew what Jamie had done to you, but I was in the grip of something powerful in those days. Even when he was in prison, I believed him. Or perhaps I should say that I <u>believed</u> I believed him, if that makes sense. I didn't have the space to doubt: you have only seen their steely side, but unless you have experienced it, you can't imagine the warmth of a Balcombe welcome. Jamie is his mother's golden boy; her faith in him went a long way to persuading me.

They set me up in a flat, helped me get it ready for his release, and although he kept up the show for the first few months, it wasn't long before he began to take what he wanted, whenever he wanted it. I don't have to spell it out, but let me only say that you and I have more in common than any women should. By then, I felt that I was in too deep. I had his ring on my finger; I'd put my name to his campaign. And there would always be charm, and presents – jewellery, clothes, perfume – in the aftermath. Never apology, though, for to apologise is to admit.

You might have children now, and if so you'll understand why I stayed so long. Material provision does come into it, especially in the early days, but he was a good father too, in so far as he never hurt me in front of them.

But there is always a turning point, and mine came when he did it again. Not to me; or, not just to me.

Let me start at the beginning, or this other girl's beginning. Jamie's firm offers a very prestigious internship for undergraduates; it's well paid, and the interns get more construction experience than in any rivals in the sector. I suppose you need to know the industry from the inside to understand what belonging to the Balcombe group means to a young architecture student. It's a huge deal. Last year, the intern was young, female and pretty, and Jamie became her mentor, giving her unprecedented attention, although she wasn't to know this wasn't standard. She was twenty-one at the time but a very naive twenty-one, God knows I know these young women exist, and she didn't realise he had designs on her until it was too late. The circumstances were horribly familiar, Beth: he made a pass one night when they were both working late, she rejected him, he brushed the whole thing off and she put it down to a flattering, if embarrassing, incident. Later, she told me that it even made her think that his conviction couldn't be sound; surely if he was really a rapist he would've taken the opportunity to attack her then? She was actually more relaxed after she'd turned him down than she had been before.

Two days later, he threw her to the ground in the office car park – the cynical bastard knew exactly where to do it, in the CCTV blind spot – he held her down and he raped her. He had her face down in the dirt, just like he did to you in Cornwall. It wasn't interrupted this time. She doesn't know how long it went on for, but it was daylight when he started and full night when it ended. Jamie's recompense began as soon as he had finished with her. He offered her a pay rise, glowing references, a chance to work on the new estate they're building down in the Thames Gateway. She was in such shock that she simply said yes. She took the deal.

Do you know how I know? She didn't tell me; she would never have told me. But she sat on my table at a charity ball, and this girl I'd known vaguely for a year or so, this sparky, brilliant young woman was dead behind the eyes. I wondered if it was boyfriend trouble or debt or an eating disorder or something, and thought, I'll corner her later; see if she needs help. She was not herself: reminded me of someone, but I couldn't place who. Then, later, there was dancing; all Jamie had to do was look at her and she ran, in heels, across the dancefloor and into the ladies'. My blood ran cold. I knew who she reminded me of. I'd seen that exact look in you and I'd seen it in my dressing table mirror. I caught her in the ladies' and she broke down and told me everything.

I tried to persuade her to go to the police. I told her I would support her but she knew my history, the campaigning I've done, so why should she have believed me? Instead, she looked at me like I was insane, a liar, or evil. She thought I was part of the conspiracy! That night, when I was taking my make-up off, I saw her disgust reflected in my own features.

The next day, I told Jamie that I was taking my statement off his website. Not that I was leaving him; not that I was going to run a counter-campaign. Just that quiet withdrawal of support was enough. He is presently in Erlestoke Prison for breaking my jaw in response.

I don't expect or deserve your forgiveness but I would so love one day to meet you and try to make some sense of this. I hope – there's that word again – that you will take this not as an insult but as a heartfelt apology.

Yours sincerely,
Antonia Balcombe

The last sheet of paper floats to the tiles. He did it. Jamie raped Beth. The doubt that's been weighing me down for years peels away. Relief wheels in circles above my head like an uncaged bird. Morally, I am off the hook. The tears that come now are for Antonia and the other girl. I could cry all night, but the clock is ticking.

The next page is Beth's writing again.

Ok, so. I've met Antonia a handful of times since she wrote this letter. She's nice. She's sincere. She'll have a metal plate in her jaw for the rest of her life. She's had it worse than me because she's had all those years of the bastard. Anyway – we've talked it through and this is the best way I can sum up what's been happening.

When she wrote that letter, Jamie had just gone inside. Since then, he's got worse, not better. The new girl, this intern – I don't know her name – is apparently close to telling the police, and he knows this. He now thinks Antonia, her, me and even you are in a conspiracy against him. He's lost it completely. He's talking about suing the intern girl, for fuck's sake. He's also, and this is where you come in, talking about going 'back to basics', which means, as far as we can work out, getting you and me to change our state-ments about what happened on Lizard Point. I know, I know – he's well gone. But he <u>believes</u> he's right, that's the scary thing.

It's hard to say how real this threat is – I mean, prison, and the street come to think of it, is full of nutters talking shit. I keep thinking, no, he doesn't mean it, and anyway, he can't find them, I should leave them in peace. On one level it's almost laughable; the idea that we'll suddenly turn around and go, 'Actually, Jamie, you're right! I was gagging for it after all, and Laura never even saw us! Hands up, fair cop, the whole thing was a conspiracy!' Then I talk to Antonia again, and I look at the way she can't close her teeth to bite properly any more because her jaw can't ever be set back to how it was, and how frightened she is, how she's struggling without the in-laws who've been like family to her her whole adult life. She wouldn't make this up. Then, it doesn't seem laughable at all. And I remember what he put me through, and I think about where he is, and the people he must be mixing with, and the fact that he's still a very rich man. This blend of money and connections and his own insane conviction makes me think Jamie's more dangerous now than he's ever been. And I think all this and I reason, I can't <u>not</u> tell Laura and Kit.

Antonia is worried about what Jamie might do, and so am I. With her gone, the reputation he's been trying to rebuild is in tatters again. All those people who thought he didn't do it, who thought he was innocent, they believed him because of her. Without her he can't fool them any more. It's like he hasn't got anything left to lose, especially if she persuades this girl to come forward. He wouldn't have been able to find you in the same way I did – it was only knowing you like I do that led me to you . . .

This last line pulls me away from Jamie and back to Beth. What can she mean, I only found you because I know you so well? It would be so easy, in this onslaught of new information, to lose sight of that. Or is this what she wants? Am I walking right into a ruse? I read on with renewed caution.

. . . but he's rich, and determined, and he's got fuck-all else to do. And yes, before you ask, I've tried to tell the police. They were my first port of call. I tried to talk to DS Kent, just to get her advice really, but she died a few years ago. So I had to go to my local force, and they were of limited help. They took me seriously, because of Jamie's conviction, but mainly they said they'll monitor him and let me know when he comes out. Pretty pointless, as I'm already in touch with Antonia. But they wouldn't do anything for you, even though you testified against him. I've asked them and so has Antonia, but no joy. For a start, he needs to make the threat to you directly for the police to be able to action it. Him mouthing off to Antonia doesn't count, apparently. It's classed as a third-party threat. Even if the police could offer you protection, there's another, obvious problem. Who do I tell them to guard? 'There's this couple, Kit and Laura, I think they've changed their names, or they might not even be in Britain any more for all I know, but could you just devote all your resources to finding them and putting them into a safe house?' You can imagine the response.

Anyway, I've tried to cover everything but there's bound to be something I've missed. I really hope we can meet face to face so I can fill in the gaps.

Thanks for reading.

B x

The last page is a list of numbers: mobiles and landlines with unfamiliar codes, for Beth and Antonia, and underneath in capitals

SAVE THESE IN YOUR PHONE. WE ARE BOTH HERE IF YOU NEED US.

I hesitate, reluctant to do anything Beth tells me; then I save her numbers so I know not to pick up if she somehow finds mine – I wouldn't put anything past her – and Antonia's numbers just in case all this somehow checks out. Then I fold the pages tightly, a cache of threats and riddles lying in my lap, eyes closed against the clock, giving myself over to thought. The gap that most needs filling in is one she hasn't even mentioned. There is no reference in these letters to our history of gifts and pictures, glass and fire.

Gradually my surroundings come back to me; the silent phone at my side; the TV still burbling away in the background, school-children viewing an eclipse in Scotland. The spilt drink is congealing around me, a sickly scent rising from the floor and the splashes on my clothes clotting. It is 8.15.

Numbly, I walk to the kitchen, grab a dishcloth and start to wipe the floor, my belly skimming the tiles as I kneel on all fours, eyes constantly flicking from the job in hand and looking at the door, the clock, my phone. I can't sleep here tonight. I'll go to Ling's; I've got a key, it doesn't matter whether she's in or not. If I go now, while Beth's in the pub, she won't be able to follow me there. In the bedroom I change my messy clothes for a flimsy maternity T-shirt that I can sleep in, a bobbled cardigan and elas-ticated jeans.

8.20. She'll be gone in ten minutes.

I pull on a coat that doesn't do up any more, pocket my phone and double-lock the door behind me before making the short walk along Wilbraham Road to the Salisbury. Bees buzz where my heart should be and if it wasn't for my hands across my belly I'd say I was feeling reckless. Let's have it all out: the trial, the fire, the photograph, Zambia; let's get it all in the open. The key to walking on hot coals is courage.

It's barely spring on Green Lanes, let alone summer, but the promise of heat to come is enough to bring the street to life; two little boys play football in the gridlocked street, the division between road and pavement dissolving. At the Salisbury, I put one hand on the huge door, standing on the chessboard steps and squinting through the etched glass. My legs twitch, as though they alone know I should run away from her, run to the police station. But this conversation is years overdue, and that residual fear of the police remains.

The pub is decorated in music-hall reds and golds. The high Victorian ceiling and gilded walls make a proscenium arch to frame Beth, the lonely actress, tiny at her table for two. She looks harmless; no, she looks *frightened*, which is not the same thing at all. I have to stop and remind myself that Jamie being guilty does not neutralise what Beth did. Nothing anyone can do will neutralise what she did.

47

KIT
20 March 2015

The *Princess Celeste* is the first boat to pull out of Tórshavn harbour and it's quite the event. The waving crowd look like something from a hundred years ago, the kind of image you usually see in sepia, where all the men are hatted and moustachioed, and all the women are waving lacy handkerchiefs.

Instead of hats and hankies it's all stormwear and smartphones. Among the crowd are Krista and her family, in their matching violet coats. I give them a two-handed wave and realise that I would really like to stay in contact. (Although I still have no idea who Bill is, and it seems the moment to ask has passed, now; I can see us being friends for another two decades, with me having no idea how I'm supposed to know him.) The child astride his shoulders is a toddler, and will still be young enough to play with my children, who will be about his age in 2017. It's a warming thought.

The horn sounds; the ship vibrates, then lurches once to starboard, making us all grab on to the barrier, which is immediately followed by much laughter from those who grabbed in time and swearwords from those who missed it.

The Faroes are soon out of sight. The ship cleaves a foamy gutter in the petrol blue sea. I stand at the rear deck, where I thought I saw Beth on the outward leg. Was that really only two days ago?

There is longing, too, surging in relief's wake. I'm semi-institutionalised to life at sea, but I'm getting sick of decks and rivets and railings; I don't want to sleep on a boat tonight. I haven't got the patience for tomorrow's train and Tube. What I'd really like is to teleport home now, on to my doorstep or, better still, into my bed, to curl around my sleeping wife, one hand on her belly, and fall asleep to my children's tiny arrhythmic kicks. I have travelled enough. This trip will last me for a long time. I won't leave London until the great American eclipse of 2017, and we will go there together: me, Laura and our children.

I send Laura a two-line text:

Homeward bound. Turning my phone off for the night. Love you
xxxx

We plunge into the North Sea, the bars on my phone disappearing one by one as the land mass dissolves behind us. When I am finally out of range, I slide my thumb and fade the screen to black.

48

LAURA
March 20th 2015

The barman bends over a copy of today's *Standard*, its front page a sickle sun through cloud. Beth's drinking a pint of soda water with a half-moon of lime in it.

'You came!' she says. Her legs are crossed at the knee and her right foot kicks to a tune only she can hear. I suddenly crave alcohol. Ling drank throughout both her pregnancies: never to excess, but defiantly. Red wine's good for you, isn't it? Anti-oxidants or something? Mac would know.

'Want a proper drink?' I say. Beth eyes my belly: her hesitation is fractional.

'I'd love a glass of white wine, thank you.'

Waiting at the bar, I can feel her eyes boring into my back. I get a pint of Guinness, for the iron. The barman pipes a shamrock into the foam. I lower myself into the chair opposite her, slopping the drinks on the wooden table. Ever the barmaid, she mops the spillage with a beermat.

'Congratulations,' she says quietly to my bump. 'I didn't know.'

Her face has changed. There's a gauntness around the eyes and a slope to her cheeks that speak of weight gained and lost, as though in pregnancy. I wonder if Beth is a mother, but I won't ask.

'Thanks.' I keep my voice hard. 'So. You've come to warn me about Jamie. If you've known for ages, what's the urgency? Why now, while Kit's away?'

'We're launching straight into it, are we?' She seems disappointed.

'You said yourself it's not a social call.'

'No, fair enough.' She produces more paperwork from her bag. Her nails are neat, a round red manicure. 'Jamie wasn't supposed to be out for another six months but they've brought his release date forward. He's out next week. I'd been umming and aahing for ages and when I found that out I thought, no, I've got to say something *today*.' She misinterprets my fear and says, 'Don't worry, probation will let Antonia know when he's released, and she's going to call me. Oh, that reminds me. Did you put our numbers in your phone?'

'Yes.'

She registers the grudge in my voices, breezes past it. 'Can you text me yours?'

I'm tempted to enter a false number, but then reason there's no point in keeping my mobile a secret when she knows where I live. I tap myself into her phone.

'Thanks,' says Beth. 'Ok, I'm going to Antonia's tomorrow morning after breakfast, catch up with it all. I might need to call you.'

It has the ring of truth, but I seize on something. 'Can't they tell you directly? Because of your history with him? Because of the threats he made against you?' Some knowledge comes back to me, gleaned from when I was researching the implications of Jamie's letters. 'Isn't it a breach of his original probation even to make contact with you?'

Her laugh is like spitting bitter pips. 'In theory you're right, but you're putting a bit too much faith in the probation service.' Her foot kicks faster; I want to grab it. 'I mean, it's not their fault, they're nice enough people, but they're working like ten cases a day each. He only made the threat through a third party – that's Antonia – so he hasn't *as yet* committed a crime. If he does contact me, he'll be in breach of the terms of his original sentence, so they can do him then. And he'll be straight back inside if he even rings me. Although I don't think he'll stop at a phone call.'

My pint of Guinness sits untouched on the table. 'And me and Kit have no protection at all,' I say. '*Fuck*.'

'That's one way of putting it,' agrees Beth.

I stare into my drink. The shamrock sinks as the foam settles, the picture fading fast. 'What's he actually threatening to *do* to me and Kit, though?'

Beth's foot stills but she starts to roll a beermat on its side. 'Well, his words to Antonia were, whatever it takes to make them eat their words. And we've seen what he's capable of, so . . .'

'So you come all this way to tell me that there's a psycho out there with a vendetta, but it's ok, you'll give me a ring when he's behind me with an axe?'

'I didn't *have* to tell you.' She notices her own petulance, checks it, and tries reassurance. 'But you're forgetting something. He doesn't know where you are.'

My biggest question is shuffled back to the top of the deck. 'But *you* do,' I say. 'You said that you found us in a way that Jamie Balcombe would never be able to.'

'Bit of detective work online,' she says. 'You're bloody impossible to find. Kit wasn't easy but I knew if I hung around the eclipse-chasing chatrooms for long enough he'd turn up.'

Anger at Kit's mistake again bubbles up but I have to be careful not to let it spew out and obscure the real issue at hand here.

'But how did the video lead you back *here*?'

She looks blank and I realise that we are talking at cross-purposes, even if I don't yet quite know what that means.

'It wasn't a video, it was a picture. On his Facebook page.'

'Kit's not allowed a Facebook page.' I no longer care if I come across as a shrew.

'Well, he's got one,' says Beth, apologetically. 'So, online, you basically don't exist,' she says as she flicks through apps on her phone. 'Neither of you does, or not in the normal ways, anyway.

So I had to think outside the box. I thought, there are so many eclipse fan sites on the web, there's no way he'd be able to leave those internet forums alone.' Screens are loading and moving beneath her fingers as she speaks. 'I went through all of them, looking for him. And eventually . . .' She has landed on a Facebook page called Eclipse Addicts, and she slides the phone across the table to me. 'I knew it was him because of the top.'

She's highlighted someone calling himself Shadowboy; no portrait, just a ragged burgundy T-shirt with Chile '91 emblazoned across the breast and I know that if you zoom in you'll see a tiny black rock burn on the neckline. 'That's how I knew it was him. I slept in that T-shirt, remember, the night after we played pool that time? That little burn itched me all night long.' I remember stuffing the T-shirt back in the wardrobe before he could see. Kit doesn't know that Beth knows it exists. But the fact of him having a Facebook account at all; my anger at him bubbles and builds.

'We *promised* each other, no online footprint, nothing that let us get found . . .'

'For what it's worth, he's done a very good job. His privacy settings are pretty impenetrable, and he never posts anything personal, just stuff about lenses and weather updates and a load of numbers that don't mean anything to me. But then, like, three months ago he put this up.'

She enlarges the photograph. It's Kit's equipment, laid out on his desk, the camera, every single lens, single-use lens wipes in their sterile packaging, a spare camera strap, two extra memory cards. Everything is at right angles. In the background, his map is on the wall; if you'd seen it before, if you'd destroyed it in a fire, you'd know it was him. The caption reads,

'All prepped for boys' trip to Faroes 2015. Not OCD at all.'

'You got our co-ordinates from his photo,' I say quietly. I'm going to kill him. I re-read the caption. *Boys' trip.* That's how she knew he'd left me behind.

'No.' She shakes her head. 'He must have disabled the location settings on his phone.' She zooms in closer on the photograph.

On the very edge of his desk, so far to the left that it's only just in shot, there's a paper cup in a brown sleeve. The picture is of such high resolution that it's easy to zoom in and read the logo: Bean/Bone, N7. And then, she zooms in to something else; if you're attentive, Alexandra Palace, the very tip of it, is reflected in the black mirror of his computer screen. She's analysed this picture forensically. It's the same dogged determination with which she tracked down his camera lens, my favourite candles. I shift back in the hard chair.

'You found the house from that?'

'No,' she said. 'The house was harder. I've walked quite a long way today, trying to narrow it down.'

'Go on,' I say, even though the tension is starting to weary me. 'Enlighten me.'

'You're sure?' Everything about her screams reluctance as she swipes and slides to another page. Shadowboy has contributed to part of a discussion about stargazing in an urban environment.

Bastard neighbours and their building work – left their sodding floodlight on all night. Stars invisible. Furious about missing Pluto's alignment.

'I walked all the streets for about three miles around the shop and looked at all the houses with scaffolding.'

'That must have taken you all day.'

'It did,' she says wearily. 'There's only one with a floodlight on it. And then I waited till I saw you in the window.'

I picture Beth, footsore and determined, waiting outside my house for the street to clear, Kit's carelessness painting an arrow over the roof, a black cross on the door. 'How *could* he?'

'It's only a Facebook account,' says Beth. 'Everyone's got one.' I shake my head; she can't understand. Kit swore to me that he would never do this. Betrayal is subjective; proportionate to trust.

'Don't be too hard on him. I wouldn't have been able to find you otherwise!'

'That was kind of the point,' I say. The silence throbs between us. 'I've lived in fear for fifteen years, you know. I went to a psychotherapist. I had this disorder where I . . .' I pull my sleeve up so she can see the lacerations where my scratching broke the skin. 'Whenever we go on holiday I have to make the hotel staff do a full fire drill with me before I can close the door behind me.'

'God, Laura, that's awful. You poor thing.'

She tries to put her hand over mine; I snatch it away. 'Are you taking the *piss*?'

'Laura, don't be like this.' Her voice is smooth but her right eyelid flickers. 'I'm trying to help you.'

I've forgotten about Jamie; he could be behind me with a machete and I'd be too lost in furious memory. 'Seriously, what were you expecting? Oh hi, Beth, you tried to burn me alive, but it's all water under the bridge now! Let's have a nice bottle of wine and a chat, shall we?'

Her pupils splash wide, black ink from a dropper. 'Hang on, *what*? I . . . I couldn't – how could you even *think* that?' Now her hand is on my wrist. My skin flames at her touch; scar tissue, like muscle, has memory. 'If this is your idea of a joke, it's pretty fucked-up.'

'Oh, Beth, don't do this,' I say. 'It's embarrassing for both of us.'

She's trembling with the effort of self-control. 'No, let's do this, actually, let's have it out.'

She asked for it. I start with something she *can't* deny. 'Ok then. How do you explain Zambia? How do you explain stalking us in *Turkey*?' Her mouth falls open; she wasn't expecting that. 'Someone filmed the festival, you were in the background with a photo of us, asking people if they'd seen us.'

We've been speaking in murmurs until now; her voice cracks wide.

'Why do you *think* I went to Zambia?' The barman, sensing a catfight, looks our way. Beth drops her voice. 'To find out why you would *leave* me that way, just when I had a chance to show you some of the support you showed me?'

I let out a not-laugh. 'I can't believe I'm hearing this.'

Beth puts her hand over her heart in a way that's presumably meant to show honesty. 'You meant the world to me, Laura. You rescued me, I *loved* you. You and Kit got me through the worst thing that had ever happened to me, and then you . . .' she mimes the pulling away of a rug. 'You broke my heart.'

The words trigger a memory. 'That's *exactly* the phrase you used about Tess before you carved up her tyres,' I say. Beth has no comeback. The hand on her chest falls limply on to the table.

'Look,' I say, 'I know you couldn't help it. Jamie *made* you this way. I can see that now. I mean, I feel sorry for you. But that doesn't change what happened.'

We stare at each other; there's a world of something going on there, but I don't know what. I keep quiet, giving her space. An admission now will restore some of my faith in her. There's a tiny shake of her head, then she visibly swallows her words.

'You don't know what you're saying.' There's broken glass and fire in her voice. 'Actually, you know what, Laura? Think what you like. Think what you fucking well like. I've done what I came to do.'

She pushes her chair back, and gets to her feet.

'I'm not going home, so don't get any ideas,' I tell her.

'Oh, don't worry. I've been insulted enough for one day.' She pauses, winds her scarf tight around her neck, and gives me a look that slows my blood. 'You can't say you weren't warned.'

As she stares me out, the fire comes back to me in all its searing horror. I've been an idiot trying to reason with her. I give the table an ungainly push, then wobble to my feet. Still only halfway into my coat, I shove through the heavy double doors and on to the blare and lights of Green Lanes.

It's not a good idea to run when six months pregnant with twins. Even with one hand on my belly for support – I need the other for balance – I can feel my pelvic floor give out with every heavy step. I cut the wrong way through Harringay Passage just in case she's after me. By the time I get to Ling's I'm a sweaty, panting mess.

'Laura!' says Juno. She's old enough to know something's wrong but not old enough to cope with it. Her lip wobbles. 'What's up, is it the babies? Come in.' She calls over her shoulder, 'Muuuum!'

Ling charges down the stairs, two at a time.

'Oh my God, Laura' she says, throwing an arm around my shoulder and steering me in. 'I got a missed call, I was just about to ring you back. Do we need to go back to hospital?'

I don't even consider the truth.

'Panic attack,' I say. 'I couldn't be on my own.'

She takes me at face value. Why wouldn't she?

'You can have my bed.'

Ling's bedroom has clean sheets, fresh flowers, white walls on to which I mentally project everything I think I know. The threat of Jamie feels real – the sentiment is familiar from his letters – but distant, like a figure on the horizon vivid only when viewed through binoculars. My instinct is to run again, start another new life, but we are tethered now, not least by hospital appointments and work and mortgage. We are embedded here. Fury at Kit rises to the fore again. I actually twitch with the desire to punch him. How could he put us at risk like this? How could he lie to me? It must be one o'clock in the morning when anger trumps restraint and I fire off a nasty text message.

> *Beth came to the house – thanks to you. She told me what you did. How could you, Kit? I'm staying at Ling's tonight but we need to talk, big time, when you get home.*

Lashing out is a huge release. Isn't part of the marriage contract to take shit off each other from time to time? We'll have our row tomorrow and then we'll make up, like we always do.

I can't deal with Beth, let alone Jamie, on my own.

49

KIT
21 March 2015

'I'm so tired,' I moan, as we stand on the bridge, coffees in hand, waiting for the first sight of land. The sky is white, the morning light is cold silver and the North Sea is a uniform dolphin grey.

'I slept like a log,' says Richard cheerfully.

He doesn't have to tell me that. I was awake for every snore and fart. I would have thought that after the relief of Beth not catching up with me, and with the pre-eclipse tension having dribbled away into anti-climax, I would have slept like a log too, but instead I tossed and turned in my narrow bunk, checking my watch through the night and plunging further into wakefulness every time.

A murmuring starts up as shapes form on the horizon. Even through its swathe of low cloud, the Northumberland coastline is an almost vulgar green after the rocky black Faroes. It feels good to see England and my pulse quickens in anticipation of home and Laura. We close in on the cranes and gantries of Newcastle dock and I wonder if it's worth hurrying to catch the ten fifteen to King's Cross or, I should just relax and enjoy the ride. Either way, I should be home by two, three at the latest.

As the famous bridges across the Tyne come into focus through the mist, my phone comes back to life, buzzing its presence in my pocket. I check the message from Laura, sent just before midnight last night.

*Beth came to the house – thanks to you. She told me what you
did. How could you, Kit? I'm staying at Ling's tonight but we
need to talk, big time, when you get home.*

The shock, the sharp vertigo tilt, is so physical that for a moment
I think the ship has struck something, and I don't stifle my cry in
time.

'You all right?' says Richard in concern.

'Must be sea-sickness,' I try to say, but a kind of slur comes out.
'I'm just going to sit down. I'll be fine, honest.'

'You look like shit. You were all right ten seconds ago. Shout if
you need me?'

I nod, then collapse on to a bench. Even sitting down, my pulse
is skipping like I've just sprinted. How the hell – why would she
– and why now?; what has Beth *said*? The woman on deck, the
woman drinking beer in the photograph, Krista were shapeshift-
ers of my own making. I had been so busy superimposing Beth's
face over all these others, I had never conceived that she might go
to London.

I re-read the message even though I've already memorised it.

She told me what you did.

Everything I've got left is going into not screaming. Fifteen years;
fifteen years I've been protecting Laura from a truth so awful and
painful I don't even let myself think about it. I realise, as the boat
crests a wave, that I have been wrong to think of time as linear; of
my life moving in one direction away from one event. It's a twisted
circle of ribbon, a Mobius strip. The years fold in on themselves,
until I am back in Cornwall, my cold hands moving over hot skin,
gold paint coming off on my fingers in a reverse Midas touch, and
the only fixed point in my history seems to be the night I spent
with Beth.

50

KIT
9 August 1999

With forty-eight hours to go before totality, Ling put into words what we were all feeling. 'This festival is going to go down in history as one of the most expensive flops ever.'

She was right. Tall, fluttering flagpoles, designed to act as markers in the throng, were spaced out in an empty field. It looked less like a thriving hub of counter-culture than a golf course on to which some ill-equipped, inappropriately dressed people had mistakenly wandered. Our little stall was doing ok, mainly because the tent was a bubble of trapped warmth from the generator and the urn. Our first customer of the day was still here, two hours into a nice drunken stupor, greasy dreadlocks spread like tentacles across the mats.

'This weather can do one,' said Mac. He glared at me and Ling like it was our fault. We clicked as a foursome but with Laura missing, it didn't work any more; I was back to the old days, playing gooseberry to Mac and his woman of the week. The three of us had been niggling at each other all day, taking out our various disappointments on each other: Mac was angry about losing money; I was pissed off with the weather; Ling was finally coming to understand that Mac was not so much a *really fun bloke* as a pathological addict. Since we'd arrived in Cornwall, he had been either drunk or high, hungover or coming down. Tiny, delicate

Ling's doomed solution was to try to match his appetites. That morning, they were both drinking tea laced with whisky. After half a dozen cups of chai, I was completely wired and pissing cinnamon. Our speakers were throwing out gentle ambient music but I had one earphone hooked into a pocket radio, tuning in twice an hour for the weather forecast.

'We're doing better business than lots of the others,' I said, although that wasn't hard. Jon, the bloke in the organic burrito van, was virtually in tears as his ingredients mouldered and his staff idled behind the counter.

'It's your bloody fault,' said Mac.

'*What?*'

'You and your bloody pie-charts or whatever, working out the sun, making me come to Cornwall. We should've gone to Turkey like I wanted to. Just put it on the credit card.'

'That is exactly the *opposite* of what you suggested,' I said. 'If you remember, you were the one who wanted to stay in England. That's why I spent a whole week doing those sodding weather charts for you.'

'You just wanted to look right-on in front of Laura. She got all excited about the West Country, and all this Fairtrade shit everywhere, and you just went with it.'

'This has got nothing to do with Laura and everything to do with the Gulf Stream,' I said. 'Cornwall ought to be sunny. It's not. There's nothing I can do about it.'

He was like a dog with a scent. 'You know what your problem is?'

I braced myself; he had so many to choose from.

'You're pussy-whipped. Under the thumb. Now you've finally found someone to shag you, you don't care about anything else.'

I felt myself blanch, even though Ling took the vicious non sequitur as a joke. Her laugh was the warm, spontaneous kind that comes of genuine amusement rather than the awkward, pitiful one of someone who's just heard an awkward truth spoken as jest.

'Fuck off!' I mouthed at Mac when Ling's back was turned. He held up his palms in an apology that was only partly feigned; even he could tell he'd gone too far. I turned my back on him, shaking at the near miss. Mac knew, was the *only* one who knew, that Laura was the only girl I'd slept with. The millstone of my twenty-one years of virginity had finally been unhooked from my neck that clumsy night in his freezing flat. It nearly makes me laugh, now, to think that once *this* was my deepest, darkest secret. I lie awake at night and *long* to be a 21-year-old virgin. He never missed a chance to tease me about it – 'Trust you to settle down with the first girl who shows you her tits' – but always in private; it was more amusing to him to dangle the secret over my head than to tell it, and spend the bullet. This was the closest he'd come to blurting it, and in front of Ling, too. I wanted to belt him. But rather than say anything that might lead us further down this path, I settled for silence and an evil stare that bounced off him like a broken satellite receiver; he'd already moved on to the next thing.

'If we're not going to make any money, we could at least liven things up,' he said. His hand was already in his pocket, the decision made. My spirits sank as he pulled a strip of perforated paper out of his pocket, marked with the suits of cards: diamonds, spades, hearts and clubs. He tore off a little paper square and placed it on Ling's tongue.

'No thanks,' I said, when he offered me one, a doll's postage stamp, the size of his fingertip, marked with a diamond. I would never have touched LSD with a bargepole. I'd witnessed some bad trips over the years and if that's what they look like from the outside, I was fucked if I wanted to experience one from the inside.

'You are *such* an old fart.'

'If you say so.' Mac thought anyone who spent more than twenty-four hours sober or straight was a prude. If this had once been true of me, it had all changed since meeting Laura. Splitting an E with her brought us even closer; it let us crawl around inside each other's minds as easily as we did over each other's bodies. The problem wasn't the drugs but Mac. There was no shared experience to chase with him. Whatever the stimulant, it was always *his* trip, and he only wanted others along for the ride to justify his own excess.

He rolled his eyes. 'Ok, you can take one for the team,' he said, his tone changing completely. 'I'll make us all another tea.'

'Not for me,' I said, but he dropped a bag into a mug anyway with a smile so sweet that my suspicions were immediately roused. He's going to spike my drink, I thought. I toyed with the idea of accepting it, then throwing it back in his eyes. I checked my watch; I had about twenty-four hours until Laura arrived.

'I tell you what,' I said. 'I can man the stall on my own. You go and have your fun.'

I saw Mac's hand hover over the mug like a poisoner in an Agatha Christie film. Then he put the tab back in his pocket.

'Thanks, bro.'

I watched them go, willing him to have a horrific trip. I hoped that he saw Ling's face melt so that her skull shone through.

There was enough work to make the time pass quickly. The cold sun set and the music was turned up, the boom-boom of the bass plucking at my diaphragm. I sympathy-bought an enchilada from Burrito Jon and ate it listening to the shipping forecast on Radio 4 and brooding interchangeably about the weather, my brother, and my pathetic sexual history.

All eclipses matter but this one seemed more important to me than any before. I wanted Laura to experience the full wonder of the phenomenon. She had shown me so much; this was my chance to let *her* into *my* world, and it had to be perfect.

At midnight, all the stages closed and the campfires began. At one, as I was thinking about winding down, our greasy friend from earlier came staggering towards the stall. Bloodshot eyeballs framed microdot pupils.

'Sorry,' I said. 'I'm shutting up shop . . . mate.'

We didn't grow up in the kind of house where you call people *mate*, and I've never been able to carry it off.

'Do us a tea,' ordered the crusty. 'Builders, nothing wanky.' He recognised in me the failings that so frustrated Mac; my obedience only confirmed them. I handed him the mug. A bowl of (fashionable, brown, irregular) sugar lumps sat on the counter. He stuck his whole filthy fist in there and ate a handful.

'Please don't do that,' I said. 'We'll have to throw them away.' He laughed, his teeth the same colour as the brown sugar.

'Cheers,' he said, and walked off, taking the yellow mug with him, which I gave up for lost, along with the bowl of sugar I tipped into the bin and the pound I ought to have charged him. 'Prick!' he shouted over his shoulder, by way of a farewell.

Feeling about an inch tall, I mixed boiling water from the urn with cold water from the tap into the washing-up bowl to clean the mugs, and was five minutes into the chore when I realised someone was watching me.

'Is this yours?' A yellow mug hovered at chest height; behind it stood a girl with black curly hair, white skin and small, dark features pinched tightly in the middle of a heart-shaped face. 'Some minging old crusty just threw it across the field. Hit me in the leg. I don't know why he's being such a knob. Something to do with the eclipse. Confluence of heavenly bodies, probably. Mercury's in retrograde. Or ley lines.'

'No, I think it's that there's some really strong acid on site,' I said. She laughed, steam curling from her mouth, and handed me the mug. I dipped it into the sudsy water.

'I'll have a cup of chai, please. No sugar but loads of milk.'

So much for shutting up shop. 'Coming right up,' I said.

'That's the problem with hippies,' she said, as the teabag steeped. 'Always the exact opposite of what they purport to be. Peace and love my arse. Ok, not all of them. I'm sure you're a beacon of sincerity. But some of the most violent, lazy people I've ever met have had a CND sign hanging around their necks or that Hindu tattoo they all get.' She could have been describing Mac. I experienced the lightbulb pop of a long-known truth finally expressed.

'Have you done the whole circuit this year?' she said. 'I haven't seen you before.'

'The circuit?'

'Working the festivals. That's what I'm doing. Cash-in-hand work in the holidays. I've gone to so many over the last few summers that I'm starting to recognise people. That bloke on the burrito stall over there, I'm sure he was at Phoenix the other year.'

'Possibly,' I said, only then seeing that Burrito Jon had locked up for the night. 'This is our first go at anything like this.' It's my brother's idea really. We just wanted a way to see the eclipse that would pay for itself.'

'And has it?'

'We've lost money by coming.'

'No wonder I can't pick up any work. There's a glut of labour. A skewed job market. I'm a hopeless victim of capitalism.' She laughed again. 'This is *good* chai.'

I smiled. 'Is it strange, going to a festival on your own?'

'I wouldn't have chosen it that way. I can usually persuade a friend to come, but August is funny this year because most of my friends from home have gone to work in Ibiza for the summer. There's a couple of people I know watching the eclipse in Devon, I probably should've gone with them.' A breeze shook the trees, shivering the wind chimes. 'It's weird here, isn't it? There's something almost sinister about it. Not just the lack of people. The way it's laid out. Usually it's just one massive field, or fields with a bit of hedgerow in, but here it's proper woodsy. There's great clumps of trees. You wouldn't be surprised to see an ogre or Little Red Riding Hood.' She looked like something from a fable herself, her skin an unearthly white. 'I'm Beth, by the way.'

'Kit.' The hand I shook was soft, like she'd been soaking it in honey, so different from Laura's long, slim fingers I almost gasped; there was no ignoring the striplight flicker of desire and I pulled away, cutting the dangerous current.

'Is it just you and your brother?' She asked the question straight, but my hand still tingled where she'd touched me. I was suddenly, acutely aware of how isolated we were. I saw myself as though in an old-fashioned film, an angel at my right shoulder to represent my conscience and a devil at my left, a little red embodiment of my animal self. *Tell her about Laura*, whispered the angel. *Tell her about Laura, now.*

I swerved the question. 'He's tripping with his girlfriend. Better to be alone than stuck with someone who's hallucinating.'

Beth grimaced. 'Yeah, acid's not my poison either. Not since Glastonbury '94 and the burning crosses coming at me across the field.' She shuddered. 'Festivals aren't the right places to lose your bearings.' She drummed her fingers on her mug. 'So what do you do when you're not doing this?'

'Just left Oxford. I'm about to start my doctorate in astrophysics.'

'I read the other day that religion's on the up in physicists,' she said. 'Apparently all the other scientists are atheists. But physicists, the ones who spend real time contemplating the hugeness of the universe, are more likely to say they believe in God than almost any other discipline. I thought that was really interesting.' I don't know what my face was doing but she smiled again. 'I'm sorry, I'm talking too much. It's just that this is the first intelligent conversation I've had in about two days.'

'Me too,' I said, and meant it. Talking to Beth was effortless. She got all my jokes, I got all hers. We swapped travel stories. I told her everything about myself except for the important thing. *Tell her*, urged the angel. *Tell her about Laura. Let her know you've got a girlfriend.* The devil simply leaned against his pitchfork and grinned. I looked up; still no stars.

'It doesn't look good for tomorrow, does it?' Beth said.

'Cloud cover right across the West Country,' I said. 'Still, we could get a break. Strong winds, you never know.'

'Speaking of which, it's really bloody cold now,' she said. 'Does this thing have an indoors? I want to keep talking but I'm turning blue.'

'Sure.' My voice, I noticed, had regressed by years; I was a squeaking adolescent trying to chat up girls on a beach. 'I might shut the tent up now anyway.' Beth watched me turn the sign to closed and zip the tent behind me.

That was when I learned my lesson about the relationship between contemplation and action. I was thinking about the logistics: the warm pocket of air in the tent; where the clean bedding was; her underwear, the buckle on my belt. The moment you think about an act in terms of how, you are already halfway to doing it.

The chill-out area, with its twinkling lights and Persian carpets, was a grubby harem. With Laura it had always been about love, but desire when laced with transgression has twice its usual pull. I thought to myself, with the unimpeachable logic of the insanely horny, that I would do this once, and then I would go back to Laura. I even reasoned – in as much as reason came into it – that I had simply deferred the women, the new bodies, the one-night stands, that were every young man's due.

A red sleeping bag in one corner was held together by a belt.

'This is clean,' I said, and unclipped it. It rolled out like a red carpet. Beth sat at one end of it; I at the other.

'So,' she said, and smiled in slow motion.

If I'd known what I was starting, would I still have done it? I crawled across the floor to her and kissed her. She tasted of spiced tea and faintly of woodsmoke. 'You're lovely,' she said. We undressed each other, occasionally yelping as freezing fingers brushed against warm skin. Her body was covered in gold paint, dancing suns that I smeared across her breastbone. I smoothed the angel wings on her back, as if to stop her taking flight. Beth was a yielding softness that threatened to go on forever until suddenly I was locked inside her. She moved slowly, her eyes and mouth fixed on mine. If she hadn't been a stranger I would have used a phrase Laura hated and said we were making love. When I was getting close, she held me still and looked into my eyes. 'You're lovely,' she repeated, but she wasn't smiling this time. I buried my face in her hair as I came, and even the devil on my shoulder turned his back in disgust.

51

KIT
10 August 1999

I woke at dawn after a couple of hours' sleep, still naked, my muscles cramping in the cold. Guilt immediately closed over my head, shutting out everything else. I pictured Laura's face if she found out; I thought how I would feel if she did this to me, and my guts twisted. How did Mac do this time after time? Why hadn't he told me that there were consequences, immediate and visceral? Why hadn't he warned me about the fear? Probably because he'd never had what I had to lose.

Beth was asleep, buried under my clothes and hers, breasts pressed together between milky arms, a rasp of stubble rash across her neck, the last of the gold paint a shimmer on her skin. Tattooed feathers tickled her shoulders. Freeze-frame images of her face in delight came back to me; beneath the guilt was a dark undercurrent of validation: I squeezed my fists to force it away. Without my warmth, Beth began to shiver, and I watched her eyes flicker open. A grain of sleep clogged her left tear-duct. I beat the urge to wipe it away.

'Hey,' she said, propping herself up on her elbows. 'I'm in the right place for a cup of tea.'

I could feel myself tighten. Once more, said my devil. Once more before the sun comes up. It's still last night. But I overrode it, and pulled on my jeans with what I hoped was an air of finality.

'Beth,' I said. She picked up on the weight I gave her name, and pulled the sleeping bag tight around her, narrowing her eyes.

'This doesn't sound good.'

'There's no good way to say this. I've got a girlfriend.'

There was a second's delay, then, where I might have expected recoil, Beth leaned in close. 'Fuck. You,' she said. She went through an awkward performance of trying to get dressed, but the clothes had lost their easy fluidity of the night before, and she struggled awkwardly into a bra that cut her wings in half, and then forced goosefleshed arms into the sleeves of her top.

'This,' she turned to face me and waved a hand over the sleeping bag, over me, 'is *not my style*. You've made me into that woman. Who sleeps with someone else's boyfriend. You utter shit.' She gave me a little shove in the chest that I deserved. 'Actually, no, it's more than that. There was something there. A connection. I'm not wrong, am I?'

She looked as though she was about to cry.

I was honestly stunned. I had assumed, since everyone else seemed to have been bed-hopping without consequence since their mid-teens, that the need for a deeper connection was a weakness peculiar to me. It hadn't occurred to me that Beth would have taken it *seriously*.

'I'm sorry,' I said feebly. I wanted to tell her that she hadn't been wrong, that it had been beautiful, but I knew I was already in damage-limitation territory.

'I'm sorry too.' She bent down to angrily lace a silver shoe. 'I don't like being taken for a . . . I don't know, what did you take me for? Actually, I'd rather you didn't answer that. Thanks for ruining my festival.'

She had her clothes on and the tent unzipped while I was still barefoot in jeans. I followed her into the silvered field. Freezing grass spiked between my toes. I begged her for a mercy I didn't deserve. 'Beth!' I shouted. 'Please don't leave it like this!'

But she was gone. The woods that she found so sinister had swallowed her whole. In the sky, clouds raked the ashes of dawn, reminding me why I was here. The coming eclipse seemed diminished by what I had done.

The loose tent flap slapped slowly against the wall in sarcastic applause. Back inside, I put my boots on and gave my clothes a forensic examination, extracting a single curly black hair from the fibres of my sweater. The red sleeping bag bore a faint wet silvery trail where we had been sleeping, the fever of last night reduced to a tawdry little stain. I couldn't leave it there. I rolled it up again, ready to throw it in the back of the van. I felt like I was clearing up after someone else, someone I didn't want to know. I felt like I was laundering a crime scene. I tucked the festering sleeping bag under my arm and walked back to our little camp. A few fires smouldered here and there and I thought about throwing the stained bedding on, but I knew it'd go up with a woof and a flare, attracting attention I couldn't handle. Instead, I threw it into the back of the van, where I couldn't see or smell it.

There were no sounds coming from the red tent. I unzipped the green one. Our sleeping bags were there, fastened together. The pillow I'd brought down for Laura smelled of her hair, clear as if she were sleeping there, and it seemed to conjure her face for me, but instead of picturing her smile I saw twisted hate. With a judder I understood what it would be like if the things everyone admired about Laura – her cleverness, her principles – were turned on me. She would know in one look what I was. She would leave me in a heartbeat.

It was not just her hurt but her fury that I feared.

I lay back on my bedroll, queasy morning light filtering through the canvas. The walls of my tent contracted and expanded with the breeze like a giant lung. I knew with utter certainty that I would never sleep again.

Mac woke me up what felt like ten seconds later, sticking his head through the zip. His eyes bulged like a cartoon character's and his tongue was green. I checked my watch: 10 a.m.

'Kit,' he croaked. Twenty years of his goading balled inside me; a boast crouched on my tongue. It felt worth it, in those swimming seconds between sleep and consciousness and then, as my focus sharpened, quickly revealed itself to be the second worst idea I'd ever had. I gathered myself just in time.

'I didn't expect you up so early,' I said. The sarcasm was lost on him.

'Am I fuck. We're just going to bed now.' His breath was rank with old tobacco. 'We went down to the clifftops. We got some *really* strong visuals. Lasers in the sky. Listen. Can you open up for us, catch the lunchtime crowd?'

'You're joking,' I said. 'I did a double shift yesterday.'

'Please, Kit,' he wheedled. 'We'll do the late one tonight. I'm going to actually physically die if I don't sleep.' I glared at him. 'We'll work the eclipse tomorrow,' he said.

'Go on.'

I didn't go straight to the stall. I felt sheened all over in Beth, and would have paid fifty pounds, never mind five, to stand under hot water and scrub. In the tiny, mud-flecked bathroom up at the farmhouse, I turned the water up so high my skin went scarlet. I scrubbed until the last smear of gold was gone. On the way back, I bought a nondescript khaki top with a hood that covered most of my face, and only then moved with ease through the crowd.

At last the site was beginning to fill up. Burrito Jon was playing loud mariachi band music to drum up custom and it had brought people to our section of the festival. At the tea stall, I took seventy pounds for a morning's work and pocketed ten for myself to spite Mac. From the moment I fired up the generator I was on full alert, expecting Beth to turn up. I didn't see her until lunchtime. She'd changed into some weird purple flares and had her hair wrapped in a tatty brown towel. I wouldn't have noticed her at all but her stillness was conspicuous in the milling crowd. When our eyes clicked together, she turned away. It was only later I realised she hadn't been after me at all. She had wanted a look at Laura.

Laura walked across the field. I nearly dropped the bag of rubbish I was carrying. Because of her job interview, she'd done something different to her hair, making it hang straight and silky rather than the wiggling waves I was used to. I flashed forward into our future, Laura putting her keys on the table and kicking off work shoes, and me closing the laptop for the night. This unambitious domestic fantasy now seemed like all I wanted from my life, so why did I get the urge to fall at her feet and confess everything? I kissed Laura and tucked her hair behind one ear, because that was the kind of thing I normally did.

'How'd the interview go?'

'Ok, I think. We'll see.' I could tell by the way she searched me, her eyes cutting star shapes on my face, that she knew something was wrong. She tried to tease me out of myself, little kisses on my ears, her hands around my waist, but I only shrank away. For the first time I understood the compulsion people get, at the edge of a cliff, to throw themselves off. I dredged up small talk about the showers in the farmhouse and the weather forecast, and every forced word was a little hell.

'You never know,' she said. 'The weather forecasters get it wrong all the time.' Her assumption almost made me glad of the cloud.

I'd always scorned the idea of extra-sensory perception but I swear at that moment I felt Beth behind me, and I turned slowly away from Laura to see her leaning against a tree, looking for all the world like just another hippy soaking up the vibe. I shook my head at her, and in response she made a tiny backwards nod that I interpreted as a summons.

Tapping into my newly emerging creative streak, I swivelled to look at the perfectly functioning tea urn. 'Oh, what's going on here?' I twiddled the temperature knob as Beth slunk behind the tent. 'It's buggered again, there's a loose connection round the back. You stay here, have your drink, while I fix this.' I kissed the top of Laura's head again.

The trees at the back of the tent dripped with wind chimes that jangled like scraped nerves. I had a sudden urge to round up every wind chime ever made and bash them all flat with a hammer.

Beth's hair had dried in long black snakes and thread veins burned scarlet against the greens and whites of her eyes. There was intimacy between us but she was a stranger, and it didn't seem right that the two states could co-exist.

'That her?' Her arms were folded but her foot was repeatedly scuffing the forest floor, gouging a trench in the dirt.

I was indignant. 'Of course it's her. I don't go around . . .' but I trailed off. Why should Beth believe me, going on available evidence? 'I've never been unfaithful to her apart from you.'

She snarled a bitter laugh. 'You're saying I'm special? You're saying I should be *flattered*?' That was, in fact, exactly what I meant; it had sounded better in my head. 'No – I don't know – I'm just saying, please don't say anything to Laura. I'm so sorry I wasn't straight with you, but this isn't her fault, this would break her heart.' My legs buckled with a primitive compulsion to drop to my knees.

A sudden gust of wind shook the trees around us; leaves roared like the sea and bells rang out of time.

Beth let her arms drop to her sides. 'Are you in love with her?'

For twenty-one years women had ignored me. Now that I apparently had worth, I wished it away. 'Yes,' I said, sensing the need for honesty. 'It's everything.'

Beth rocked from side to side, as though she were physically weighing her options. 'I've got to say it looks convincing, from the outside, at a glance. Hard to say without getting her take on it all.'

Everything inside me went loose. '*Please* don't say anything, I'm begging you.'

'I don't need to.' Her words round, controlled, bulging with unshed tears. 'If it's as good as you say it is, I'm not going to wreck her life just because you can't keep it in your pants. And if you're the arsehole that I think you are, she'll find out sooner or later.'

She kept her word; that is, the dreaded confrontation failed to materialise. I saw Beth twice that day, watching us across fields, through crowds, as though by examining us from a distance she could measure what we had. But how could she, from the outside? Even I hadn't known what I had until I'd come so close to fucking it up.

52

LAURA
21 March 2015

The light from the screen hurts my eyes as I check my phone with the wariness of someone coming round from a drunken stupor. That's how last night feels: surreal, dreamlike, beyond my control. Kit hasn't responded to my angry text and I think of him, fast asleep in his bunk, unaware of the storm he's coming home to. I don't know if I want to hit him or hold him.

At seven, my phone rings, Kit's photo on the screen; he should be pulling into Newcastle round about now. I decline the call; this is a conversation we need to have face to face. Sixty seconds later the messages start.

> *Please pick up*

> *Look I know it seems like the end of the world but we can get through this*

> *Please talk to me baby*

> *I wish you hadn't found out that way but I promise I can explain, we can get over this*

> *I'm so so so sorry*

It is at least gratifying that he understands the severity of what he's done and its potential consequences. But I text back, just to shut him up.

I really don't want to do this over the phone. I'll see you at home.

While Ling dashes around trying to gather her case notes, and find a matching pair of shoes, I go over the book that Piper was supposed to read last night, and carefully forge the appropriate page in her reading diary. Juno haggles with Ling for an extra three pounds for coffee on the way home. Mother and daughter are the original immovable object and irresistible force. I've got all this to look forward to; I wonder again whether I'm carrying boys, girls, or one of each.

I catch Juno on her way out and slip a fiver into her blazer pocket. She rewards my wink with a rare kiss before disappearing, leaving an unsubtle waft of celebrity perfume in her wake. After Ling heads for the station, exclaiming at the novelty of leaving for work on time, I walk Piper the fifty yards to her school, holding her hand to cross the road and taking the old comfort and pride that strangers might think she's mine. This will be my twins' school and I notice with concern that the infant school gates have been colonised by blondes in Breton tops; the ethnic diversity you see in the upper years slowly disappearing. I disapprove of the social cleansing of my neighbourhood, the influx of yummy mummies, even as I acknowledge that, to the outsider, I'm one of them.

I let myself back into Ling's house. Wet towels carpet the bathroom floor and I find myself picking up after Juno like I did when she was a baby. I need a shower and clean clothes but home no longer stands for sanctuary. I fuss some more, straightening the pictures on Ling's walls and unloading the dishwasher until it occurs to me, as I plié inelegantly to retrieve the plates, that I've got an idiot-proof way of checking Beth's whereabouts: her landline and Antonia's programmed into my mobile. I call Beth from

Ling's house phone, withholding the number. It rings out three times and then there's a digital beep as she picks up. I cut Beth off before she's got to the first groggy syllable of hello, free to go home. Even with a direct train or clear motorways, she can't get to me before Kit.

53

KIT
11 August 1999

The eclipse itself was respite from Beth. I had a conviction, very shortly to be brutally disproved of course, that during the shadow all extraneous human activity and motivation was put into a kind of suspended animation. Even though we were clouded out, the shifting violet light on the horizon surrounded us, and with the sky restricted I felt it with my other senses in a way I never had before. And Laura was by my side, which made all the difference.

After fourth contact, we climbed down and picked our way back through the desolate car park, where vital-looking components of fairground equipment littered the ground. Back on my guard, I was too busy looking over my shoulder to notice the purse on the floor. But Laura saw it, and could no more walk past it than most people could an abandoned kitten. I took the opportunity to scour the trees around me for a pair of watching eyes, but there was no one.

Laura was gone for longer than I'd thought, and I remember the first stirrings of frustration. When I called her name there was no reply, and frustration turned to dread, or is that just something I've decided with hindsight? I re-trod her path into the caravan park, past an old bumper car and a sinister carousel horse.

The man I would later come to know as Jamie Balcombe walked backwards into me, treading hard on my toe. I could feel even in our brief collision his strength relative to mine. At my shout of protest, he leapt as though he'd been electrocuted. Laura stood, ashen, in the dark corridor between two trailers. What had he done to her?

'There's a girl.' Laura cut into my nascent horror with a trembling voice. 'I think she's been . . .' She gulped down air. 'I think she's been attacked.'

'Is she hurt?' I asked. Laura gave me a withering look. 'You know, is she hurt in a first-aid sort of way?'

I don't know why I asked. Neither of us knew first aid.

Of the woman, whose body was folded into a caravan doorway, I saw nothing but a bent and bloodied white knee. The poor, poor girl. It was appalling that anyone would assault a woman at any time, but *baffling* that it could happen during a total eclipse. It was like some throwback to the dark ages where people ranted and raved at the shadow. What weird timing, what a *waste*.

I think – and maybe this is hindsight again, because from this point on things happened so quickly that I only had time to react in retrospect – that I disliked Jamie instantly. When he told Laura to calm down, she looked to me for protection; she had never done that before.

'If you haven't done anything wrong, you haven't got anything to worry about,' I said. It was supposed to ease the rising tension but Jamie must have interpreted it as a threat. 'Would you fucking *say* something so we can all get on with our lives?' he snarled over Laura's shoulder. He looked at me as if for solidarity. I only stared. I had a camera across my shoulder but it didn't occur to me to use it; instead, I found myself trying to memorise him for the photofit, but he looked all wrong. The patchwork mugshots you saw on TV were always of square-jawed, broken-nosed thugs. It seemed inconceivable to me that the files would even contain the components of such a boyish jaw, such a smooth brow, which lowered when he realised he had no ally in me. 'Fuck this,' he said, and walked off, his pace slow but his gait far from casual. While his face was boyish, his build was not. His shoulders were twice the width of mine. I recognised the deceptively lean rower's physique; you met them all the time in Oxford. Bodies slightly overdeveloped at the expense of their brains. Already out of my depth, I sank a little farther.

'Kit, don't let him get away!' said Laura. Her hands waved wildly in the air. 'Go after him!'

It was almost funny. What was I supposed to do? Get him in a headlock? *Fight* him? As I stalked him through the maze of vans and trailers, I tried not to think about the consequences of someone like me taking on some nutcase surging with testosterone. He would pulverise me. Then I pictured Laura, remembered the trembling bloodstained knee and the two images seemed to blend together. I found that imagining Laura as the victim lit a dark fire inside me, and that I could, after all, summon the strength I needed to give chase.

When I lost sight of Jamie in the camping area I was almost relieved, but then my peripheral vision was snagged by a lone figure dancing through the tents, picking his way almost comically over the interlocking guy ropes. My heart in my boots, I followed him. Naturally I went flying at the first attempt, a slapstick pratfall over a tripwire. By the time I was on my feet, Jamie was on the far side of the field, approaching the fence of trees. There was a small but dense crowd surging past the trees and it absorbed him instantly.

Failure suspended me in the moment. In the background was the ever-present bass and a chattering crowd. The Ferris wheel creaked in rotation. Close by, birds sang their second dawn chorus, gently shushed by the leaves. Over all this my blood roared in my ears. This time, I felt no relief; only a crushing sense of failure for the girl I had let down. If I'd had a tail, it would have been tucked between my legs as I picked my way back through the tents to Laura.

I used the abandoned dodgem as a marker and eventually found Laura with her back to me, frowning into her mobile phone. The girl was still crouched in her doorway, the same pale leg extended now to display a battered silver trainer, the dip of an ankle and curve of a calf that had been hooked around my thigh days before.

My first thought was that it was a terrible coincidence but scientists know there is no such thing. Using the evidence in front of me, I came up with the only plausible hypothesis. Beth had been following me – us – and someone had either followed her, or seen her on her own, and . . . it was too horrific. That dark flame flared again, burning through the surface of my guilt. I had a suicidal urge to hold her; it passed. For a few selfless seconds, my instinct was for Beth's comfort rather than my own survival. It might not have triumphed, but it was there for a while. I cling to that knowledge.

In front of me, Laura swore at her phone, then took a few paces forward.

I did the same, dislodging a pile of tent poles someone had dumped on the ground, and Beth whipped around to see what the noise was. Our eyes met in mutual, awful comprehension. Her face was smeared with mucus, although her eyes were dry. She came here because of me, I thought. Laura would have been furious enough at my infidelity, but that it had led to *this*: any chance I'd had of being able to tell her vanished, a pinprick of light dwindling to blackness. I dropped to my knees and whispered, 'Oh, *Beth*.' She stared through me. 'Oh, you poor girl. What's he done to you?'

'Like you care.' Her voice cracked on the word.

'Of course I care, it's just . . .' I nodded at Laura's back.

She had finally found a signal on her phone, and her voice, high on the breeze, the wind in her favour, made it clear that things had escalated while I was away. 'She's traumatised, she's not really talking properly. I'd say she needs an ambulance. Can it be a WPC? Can it be a female paramedic?'

The decency that had flared in me died just as quickly. It was clear that two irreconcilable things must happen. We had to do whatever we could to help Beth. And we had to do it without revealing what I'd done.

'Beth, I'm so sorry,' I whispered, and only then realised I was not expressing sympathy for what she'd been through but apologising for what I was about to do. 'When the police get here you can't tell them about the other night.' You don't live with a woman like Laura without absorbing this kind of stuff: how the system is weighted against rape victims; the tricks clever men use. Beth carried on with her dark green stare. I couldn't even tell if she understood me. 'You know what I mean. They'll come to the wrong conclusions about you.' I was taking a risk, digging myself in even deeper. There was more now to be discovered; I knew that Laura would have been more disgusted at my cynicism than the act itself. 'I'm trying to help you,' I said to Beth. Only a tiny involuntary curl of her lip showed that she'd understood me at all. 'I wish there was something I could do.'

Twenty feet away, Laura was describing the attacker to the emergency services. I sprinted round and caught her as she was winding up the conversation. Upon seeing that I was alone, Laura narrowed her eyes at my failure. I offered no excuses; I barely trusted myself to speak.

'I'm going to sit with her till they come, see if I can get her to talk,' she said. The following seconds were thick with a concentrate version of the terror that has spread itself thinly across every subsequent day. I couldn't tell her not to without giving myself away, so I watched her disappear between the caravans and kneel down beside Beth. I pictured Laura's face when the news hit. I'd never seen her in shock or in grief but I could easily rearrange her features into scooped-out eyes and a howling, hollow mouth and knew I could not bear it. The solution came to me, gloriously simple. If Laura finds out what I've done, I'll just kill myself, I thought. I can no more live with Laura's reaction than I can live without her. She'll want me gone anyway. It was as though I'd conjured a piece of glass to break in case of emergency, and it brought a kind of peace. Crucially, though, I didn't consider a method. If contemplating logistics is a step towards action, then I can't have had the courage of my convictions. Even as I mooted suicide, self-preservation had other ideas.

Behind me, the women whispered. It was my job to look out for the police. I stood next to the broken carousel horse, with its flaking gold leaf and chipped eyeballs, with my feet apart and my arms folded in a nightclub bouncer position, braced for Laura's furious shriek.

The police were, thankfully, on us in what seemed like minutes, buzzing towards us in their black-and-yellow stripes. Laura took over at once. 'They could've sent two women,' she said under her breath, brushing past me. I knew then that whatever Beth had told her I was not part of it. I felt relief, then an emetic guilt at this relief, then bitter regret, a pattern that has repeated itself constantly in my mind ever since. Beth's continuing silence, essential as it is for both of us, is a guillotine over my neck. Although she holds the rope, she cannot let it go without it burning her own hands.

54

LAURA
21 March 2015

Wilbraham Road is as alive today as it was dead last night. There seems to be a builder's van for every car. A cement mixer churns in Ronni's front garden: good. I can scream at Kit without the neighbours overhearing anything.

I shower, rubbing stretch-mark oil in circles into my baby bump. I wash the last pink berry stains from my hair and blast it with the dryer. There's half an hour before Kit's due home. In the study, I spread before me Beth's little dossier on Jamie. I tear out the staples and cover the desk with individual pages, as if the clue to the understanding is not in their content but their formation. I'm aware of the irony that I'm planning to explain to Kit something I barely understand myself. The truth, and the threat, seem to shift with every rearrangement.

I must telephone Antonia Balcombe. I've made a living from awkward phone calls and I've trained people to do the same, yet I can't remember being more nervous about dialling a number. What would I tell trainees? That nerves only attack if you don't know what you're going to say; have a script, believe you're right, be clear about your objective. What do I need from Antonia? I list bullet points on the back of one of Beth's print-outs.

- *Corroborate what Beth said, how true is what she said about Jamie*
- *Sympathise with her/congratulate her on finding the courage to leave*
- *When is Jamie out, maybe we can cut out middleman and you can keep me in the loop yourself – Beth out of the picture*
- *Offer to meet up – she come here or me go to her?*

The final point is really a warning within a question.

- *How do you think Beth seems? Is she stable?*

I call the landline because I don't want to catch Antonia on the hop with this. The phone rings out six times and then there's a click and whirr of an old-fashioned tape-recorder answerphone, the kind that broadcasts the message to the whole room. This ties another knot in my tongue.

'This is a message for Antonia.' For the first time in years I give my old name. 'It's Laura Langrishe. From . . .' I'm about to say *the trial* but then picture Antonia's kids listening in. 'From Cornwall, from Truro. I, ah, I met up with Beth Taylor yesterday – well, you probably know that by now. I think you might have been expecting me to call. There's a lot I'd like to talk to you about.'

Should I warn her not to let Beth get too close? I chicken out. Instead, I dictate my own numbers, landline and mobile, to the machine, and then ring off, setting the phone down next to me on the computer table.

I look again at Kit's map of eclipses. His trip complete, he can now replace a red thread with a gold one, stretching in a little arc north of these islands. He's even cut the gold thread to size and set it out on top of the printer. I think about pinning it on the map, just for something to do, then think better of it. To deprive him of the ritual would be like opening a child's advent calendar. A shaft of tenderness breaks through my anger at him, confusing me further. I want his help. I want to scream at him. I want him to hold me. I want to push him down the stairs.

Downstairs, the walls and floor are clean, for the most part, but the bin is already starting to smell, a horrible cloying fruity odour that's mixed in with the dried-out bone broth cup. I pause for a moment of brief nostalgia for my blunt pre-pregnancy sense of smell, then haul the bin bag and its contents through the hall. On the front doorstep I drop the lot in the wheelie bin. The whole job takes less than four seconds. I've turned my back to the street when I feel a rush of air at my neck, displaced by the body behind it.

'Laura!' Beth barrels into me, pushing me into the hall with more force than I could ever equal. I trip on the threshold and fall forwards, leading with my belly as the Minton tiles rise up to meet me.

55

KIT
8 May 2000

On the morning of the trial, I woke up with a long, hungry intake of breath, like someone had been holding a pillow to my face. I was momentarily disoriented to find myself in a chintzy room above a pub. Laura slept – also fitfully – by my side, a twitching innocent embodiment of everything I had to lose. We were in Truro town centre but you would never have known you were in a city from the silence outside. Where were the refuse lorries? The sirens? The fights? I lay awake in the hotel room, which was so suffocatingly floral I was surprised it hadn't given me hayfever, and wished myself back in London.

I was uneasy about leaving my brother. Our father's death had been a release that bordered on anti-climax; or it would have been, had Mac not taken up the hellraiser's baton like it was a bottle the old man had pressed into his hand. I worried about him; I worried about Ling and little Juno: but most of all I worried about myself.

I was shot through with the stupid, childish longing for a magic wand. When I was young, and my world consisted of my family, my telescope, and a yard of Philip K. Dick novels, I daydreamed constantly of Kit's Adventures in Time Travel. I ran through the usual fantasies from killing Hitler to fixing the lottery, but now I knew that, given the power of time travel, I'd return to the previous August and intercept Beth on her way to Lizard Point. One momentary loss of reason had undermined my whole sanity; I took monthly health checks at the STI clinic and kept them up ridiculously, superstitiously, long after any infections I might have passed on to Laura would have showed up.

That morning, I stared through dim light at the shipwreck painting on the wall, wondering where Beth was now, and whether she had slept. She must have been reliving from the inside the sickening image that played over and over in my mind's eye, of Jamie tearing into her. The thought of it made me whimper; beside me, Laura stirred in her sleep. I put my hand on her shoulder until she stilled again.

My heart hurt with the hope that Beth would keep quiet about our night together.

I am glad that I didn't know then how impulsive she was, or it might have stopped beating altogether. No matter how hard the defence might go at her, I willed her to keep sight of how my quick thinking, in the moments after Laura found her, had made this trial viable.

I had been waiting for ten months to get the call from the CPS saying that they knew about my involvement with Beth, that she had confessed to it at the last minute and their whole prosecution would fall apart if Jamie's defence so much as blew on it. I was merely waiting for it to come out at trial.

There seemed infinite variables to consider. Would Beth break down and tell them about me under cross-examination? Had someone seen us after all? Had they found my DNA on her? Hair, skin; we had been all over each other. Semen. We had both showered the following day, but body fluids loitered internally for days. While they did not have my DNA on file, I was sure that, if they mentioned an unidentified male, I'd do something to give myself away – to Laura if not to the court. At night I dreamed that the defence team had somehow swabbed me in my sleep, and had dreams where they called me as a surprise witness, my infidelity literally under the microscope.

It was only when we were back in Cornwall that I realised finding traces of a man other than Jamie on Beth's body had implications for her, too. I knew from Laura that although they weren't supposed to get a woman on 'loose' behaviour, it didn't stop them trying.

Laura's insistence on hanging around the courtroom that first day was pure torture.

The only alleviating factor was that she appeared too wrapped up in the process to notice that I'd gone into lockdown; or, if she did, she attributed my strange behaviour to a grief that was in truth so far down my list of priorities that I wondered sometimes if I'd ever get round to it. I'm still not quite sure that I have.

I made the effort to talk about the case because saying nothing would have looked suspicious, but it was like playing tennis with live grenades. I tried to talk in generalities, about the nature of consent, but even then I couldn't seem to say the right thing. My life since that week has been lived as though with a glass of water balanced on my head. It is just about possible to walk without spilling a drop if you devote all your concentration to balance. These days, protecting Laura is something so internalised that I can do the emotional equivalent of dancing or turning cartwheels without spilling a drop, but back then everything I did was controlled and conscious and the result was a physical tension, across my face and shoulders, that fast became as much a part of my body as my head or my hands.

On the third day of the trial, when Fiona Price stood up to cross-examine me, I almost confessed on the spot to pre-empt the grilling I was convinced was coming. But I held my nerve, and it paid off when I realised her baffling questions were a set-up for Jamie's cover story about drugs in his pocket. Later in the proceedings, I was sure I'd given myself away when the barrister asked the doctor if she'd found any ejaculate when she swabbed Beth and I shot forward like I'd been catapulted from my seat. To my paranoid mind it was as good as a confession, but Laura just rolled her eyes, then turned them straight back to the evidence.

56

KIT
31 May 2000

I didn't get Beth alone until the second time she came to visit us in Clapham, and even then, privacy meant a hurried conversation while Laura showered. My dirty little secret had kicked off her silver trainers at my front door and was barefoot at my stove, slowly scrambling eggs to go with the smoked salmon and bagels she'd bought. Her straight back was turned to me. There was nothing in her posture to suggest anything like the anguish that pressed down on my shoulders, so much so that I was surprised whenever I brushed my teeth not to see a stooping zombie in the mirror. Far from freeing me up to move on with my life, the effort I'd put into keeping it together in the run-up to the trial had used up a whole year's reserves of discipline and the London life I'd been so keen to return to was falling apart. I wasn't sleeping. The undergraduates I was supposed to be mentoring hadn't seen me for weeks and I'd had a written warning from the department about my continued absence. Mum had gone from nursing a dying husband to worrying about a degenerating son. Mac had taken cash from her purse, forged cheques in her name and sold her computer. I was only lending him money because I couldn't bear to have him steal from me.

'Beth.' I kept my voice low even as Laura sang tunelessly over the whirring Vent-Axia and the patter of water on the curtain.

The irony weighed heavy on me that only illicit lovers usually understand the urgency of such snatched moments. 'I'm sure the verdict won't be overturned. There's no way they'll get an appeal past a panel of judges.' (I'd read on the internet that judges hate appeals; it calls their judgement into question and if you take that away, what have they got left? The only thing they hate more, apparently, is perjury.)

Beth cracked an egg on the side of the pan with one hand. For a second I thought she was going to ignore me, but instead she took the pan off the heat and faced me, arms folded across her chest.

'Meaning, I can move on with my life after all? Meaning, I should fuck off out of yours?' Her fury was a reminder of how carefully I must tread.

'That's not how I'd have put it,' I said, although it was exactly what I had meant. The gas ring burned blue behind her elbow: she didn't seem to notice.

'You think I'm going to tell her about us,' she said. I winced at that 'us' – I would have preferred a 'what we did', or even 'your mistake'.

'Are you?'

Before answering, Beth wiped her hands on her skirt with the weariness of an old-fashioned scullery maid. If she was toying with me, she didn't seem to be enjoying it. 'Laura's about the only friend I've got in the world right now,' she said flatly. 'Or the only one who understands about the Lizard. I can't get through this without someone to talk to. Telling her about *us* would be the surest way to have no one. And I need someone, ok?'

How could I begrudge her a friend? I just wished, for the millionth time, it didn't have to be Laura, and to wish *that* was to wish for the millionth time I'd never given Beth cause to follow me.

'I'm sorry,' I said.

'I bet you are,' she said. She attempted a brave smile; she didn't get past rueful. 'D'you know what the worst thing is? You're the only one I can have a completely honest conversation with. The only one in the world who knows the whole story. And you can't stand the sight of me.' She finally turned off the gas. 'I don't want *you* any more, if that's what you're worried about. I'm not really looking for a boyfriend right now, funnily enough.'

I chewed that over; it had barely crossed my mind. I certainly wasn't thinking of her in those terms any more. Risk might have been arousing but actual danger thoroughly wiped out that first visceral desire. I was worried more that Beth was out to *get* me than she was to *have* me.

'That's not what I'm saying,' I said. What I wanted to say was, if you don't have the self-control to think twice before turning up unannounced, before showering us with expensive, out-of-proportion gifts, before outstaying your welcome like this, if you don't possess even these basic social sensitivities, then how can I trust you with the big stuff? This was the kind of mess Laura would have been able to refine into one precise sentence, but all I could come up with was, 'You must be able to see why I'm uncomfortable. What we . . . it's bound to slip out, during one of these deep and meaningful chats you keep having. You can both really put the wine away.'

She shrugged. 'I don't know what to say to that, only that it hasn't come out yet.' Now she looked awkward. 'You know, I can see you two are solid. I told you the morning after I didn't want to be that girl who trashed someone else's relationship. I can't change what happened but I can do the next best thing by keeping my mouth shut.' She looked slowly up and down the body she'd seen all of but there was nothing suggestive about it. 'You'll just have to trust me, won't you?'

I knew, because I do it myself, what she was trying to do. Lock the problem down. Compartmentalise it. Live life untouched by it. It was hard enough for me and I was – had been – a hyper-rational, disciplined man. How could traumatised, fucked-over Beth be expected to master it? Invisible fists squeezed my lungs. It couldn't continue.

'Kit,' said Beth patiently. 'People keep worse secrets than this all the time.'

'Not me. Not people like me.' I thumped my chest to make my point; it knocked air back into me at last.

'Yes, people like you.' Her voice was shot through with steel. 'People exactly like you. Well-brought-up boys do bad things the whole time, and then they lie about it. Haven't you been paying attention?'

The burst of strength quickly gave way to tears. We were back in the courtroom; back in the field. I couldn't have pushed her any farther even if I'd known what to say.

'I'm *sorry*,' I said.

'Me too.' She turned her attention back to the pan.

The Vent-Axia ceased its droning and Laura burst from the bathroom. A plume of perfumed steam mixed with the clouds of butter that sizzled from Beth's pan.

'Smells amazing!' said Laura, as she flitted from the bathroom to the bedroom wearing only a towel, water sitting in fat droplets on her slim shoulders. The tenderness I had momentarily felt for Beth was transferred wholesale to Laura.

'Please, Beth,' I said, when we were alone again. '*Please* leave Laura alone.'

I felt that she would be in my home for ever; and I knew that I could not survive it.

'*I can't.*' Beth spoke with genuine regret, like the matter was out of her hands. I realised, with an internal plunge of sorrow, that I had no choice but to take it into mine.

August 2000

Only a year before, I had taken for granted that my wide life would only continue to expand as it had. I had a brilliant degree behind me, a stellar career ahead of me and, the impossible dream, a beautiful girl to love and to travel with. Since Cornwall the pattern had reversed and my life was narrowing to its crisis. Two images, each as terrible as the other, were my default daydream, filling times of repose as a screensaver scrolled across my laptop. The first image was Laura's face when she found out about me and Beth, my own waking adult nightmare of the storybook drawing that had terrorised her when she was little. The second picture was me, alone in this flat, with all her things gone, staring into the black hole of my future.

I think now that I was quietly having a nervous breakdown. It was a slow leaking of sanity, morality, even intelligence. The undergraduate essays I increasingly neglected to mark were frequently beyond my grasp. I would walk into rooms and forget why I was there; I'd go out to get a pint of milk and stand paralysed before the fridge in the 7–11 before coming home with bread. I started going for eight-hour walks and pretending I was at my department. Occasionally I would accompany Laura on her morning commutes, taking the Tube home again so that I could cry. These private howling sessions in our flat could last all day. They were internally painful, as though previously undiscovered tectonic plates were breaking apart to form new continents inside me. I'd keep one red eye on the clock, ready to travel back into town and accompany her home again.

I was so desperate to be rid of Beth that I retreated into boyhood fantasy again; now, time travel gave way to making Beth disappear, teleported or zapped to a parallel dimension. On a more earthbound level, perhaps she would be offered a job she couldn't refuse; in New Zealand, for example. Perhaps she would find a new friend at her job, or in her bedsit. Perhaps – and this was the most outlandish, far-fetched notion of them all – perhaps she would simply have her fill of Laura. I wanted it to be painless for Beth. I never lost sight of how much she had already suffered. All the while, their friendship deepened by the day. I'd frequently come home and find the only two women I'd ever slept with cosied up in lopsided friendship, bound in profound, female communication I could never hope to penetrate. I felt stretched, as on a rack, between my duty to see and support Mac and my desperate need to stay at home and supervise Laura and Beth.

I thought constantly about how I would issue the ultimatum to Laura, knowing that I never could. I was scared of the answer. I kept mentally replaying Laura's words: you're acting like it's an either/or situation. She had never denied it. The only way Laura would let Beth go would be if Beth herself pushed her away. And that was never going to happen.

There was a glimmer of hope when Beth gave us the photograph. Even as I stared at it, panic rising inside me, I thought: Laura's going to head for the hills after this. It was a lapse in self-censorship I couldn't have planned better myself. I thought, surely Laura won't tolerate this. If anything, I was braced for the confrontation, worried that this would be the explosion where Beth, suddenly realising she was on the verge of losing Laura anyway, blurted the truth. But the confrontation never came. Instead of being horrified at Beth's voyeurism, Laura was actually *charmed* by the picture. I'd got used to keeping myself under control, but even then I was surprised by the restraint I showed that day, agreeing with Laura as she said it was beautiful. I remember that she had her hair tied back that day and I grabbed it so tightly that if it had been any other part of her body she would have screamed in pain.

With Beth's guard so far dropped that this seemed acceptable, there was no predicting what she might do, what she might say, next. This was both my greatest fear and my motivation. I hit my tipping point the day she let slip about hitching down to Lizard Point.

'Do you have any idea,' Beth had said, 'what it's like, spending all this time with you both and having to constantly bite my tongue?'

Laura blanched; she must have picked up on my own secret terror. After Beth flounced out, I walked to the balcony, gripping the railings for balance.

'Where's she gone?' said Laura, when we both heard the street door slam.

Beth crossed the road away from the common, towards the Tube. 'On to the common, through the trees,' I said.

I waited till Laura had gone, then pocketed my wallet and took the stairs so fast it felt like flight. The traffic was against me, London motorists not wanting to give up their ten-second spurt of movement between two red lights.

I wasted valuable seconds feeding my Travelcard into the slot. By the time I was underground, Beth was on the platform, the sulphurous backdraught from an incoming train sending her scarf flying. I caught up with her as the doors opened, and grabbed her by the upper arm, soft flesh turning to iron with the shock of unexpected touch. It was the first bodily contact we'd had since that night in Cornwall.

'Wait,' I panted. 'Please, wait with me.' With a slight forward motion she tested my grip against her strength, then slackened in defeat.

We stood there in the diesel slipstream of the train as it pulled out for Edgware. For a moment we were the only people on the slender platform. The pit of the sunken train tracks gaped invitingly before us. It would be so easy to—

'Do you know how it feels lying to Laura?' she broke into my thoughts. 'Of course you do.'

She let me lead her back to the bench. When I sat down, we were both shaking.

'If she knew about us, it'd only make it worse,' I said. 'And we've kept it up for so long, now. It's not like it just happened yesterday. You've been lying to her your whole friendship. You'd break her heart if you told her. You don't want to do that any more than I do.'

'It's just so hard, living a lie. I didn't understand how exhausting it would be.'

'Beth. Are you threatening to tell her?'

A train blew in. She boarded without answering me. From the doors she said, 'I really don't know how you do it.'

Was that a threat? I didn't know. I knew only that we couldn't see her again. It was all too close to the surface now. The countdown to my destruction had been activated. I saw it in my mind's eye, red digits on a black display spinning backwards towards zero, just as time had ticked forwards at the trial. This was no longer a case of me trusting Beth to keep quiet. It was about removing ourselves from the situation, or losing Laura for ever.

I watched so many trains depart without boarding that a guard from upstairs came down to see if I was all right. I think he thought I was contemplating suicide. I think he might have been right.

That evening was the first time I cried in front of Laura; really cried, really lost it. She couldn't hide her horror in time. If I didn't do something, if I didn't think of a way out soon, her sympathy would turn to contempt and I would lose her anyway.

My eureka moment came at last when Laura was asleep and that lone candle flickered on and on. Tight with anger at the trap I'd made for us all, I wanted to pick up the candle and hurl it across the room in frustration, and that's when the idea came to me. All my life people had told me I had no imagination, but the idea for the broken glass was as vivid, sudden, and seemingly as little to do with me as an image downloading on to a computer screen.

In my defence, it never seemed *reasonable*, only necessary, and if I could have thought of another way to extricate Laura from Beth's clutches without hurting either of them, I would have done it. But I had not had a better idea. I had not had *another* idea. That buzzing neon light of stress had finally started to melt my brain. When Laura was asleep, I went downstairs, smashed the glass on the pavement outside, then carefully posted it through the letterbox. Laura would surely draw parallels between this and Beth's attack on the car and, if she didn't, there was enough now for me to guide her to that conclusion.

I spent the rest of the night knocking back one coffee after another in front of my laptop. I had all night to change my mind but the loss of reason that had come out of nowhere the night I met Beth now seemed to be an external force, guiding everything I did. Laura was, I'd noticed, newly in the habit of going down to the doormat in bare feet and, given that I was acting for her protection, I couldn't let her be injured. I waited until ten o'clock to go downstairs. There was no post on the mat, only the sooty shards of glass I had posted through the letterbox the previous evening. I chose the longest one and set its point upright.

I closed my eyes, clenched my fists and stamped down as hard as I could, my punishment starting already.

57

LAURA
21 March 2015

My elbows take the brunt of my fall. Two lightning forks of pain
shoot up my arms and fuse at my shoulder blades. The bounce of
my belly on the floor tiles is relatively soft, a sloshing inside me
the worst of it. Too stunned to scream, I roll on to my side, to find
myself eyeballing a pair of man's deck shoes in brown leather. I
let my gaze travel upwards. Red socks. Beige chinos. A flowery
shirt with pinstripes at the collar and cuffs. His head is in silhouette underneath my hall light, fine hair in a thin aureole. I don't
need to see his face to know who he is.

'It's not what it looks like,' says Jamie.

Does he know these are the exact words he used in Cornwall?

The three of us are frozen for a moment: me on the floor, him
towering over me; Beth caught between us, her hair wild across
her face and neck. I search Jamie for signs of intent. He looks the
same, but different. Thicker around the neck and slacker around
the eyes, like we all are. Prison hasn't knocked any of the rich out
of him but he's softened by the wife-bought clothes.

'Laura, I . . .' begins Beth. The look Jamie darts her way has
her scrambling to her feet just as, years ago, Antonia switched
seats in court at his nod. The sickening thought occurs to me
that they are somehow in this together. That all of yesterday –
the notes, Antonia, the pub – was a convoluted trap. I should

never have opened the door to her. I should never have come back.

'Get out of my house,' I say.

Jamie kicks the door behind him and reaches out with one hand to hook the chain in its catch. 'Come on, sit up,' he says, extending his hand. If you only had a recording of his voice to go by, you'd say he sounded concerned. I ignore his hand, put my palms on the floor and heave myself up into a sitting position, legs either side of my bump, hands checking for movement inside me. One baby kicks, then the other. My splintered joints aside, I feel . . . ok. Intact, physically at least. 'All the way up, there you go.' As I use the banister to pull myself up to my knees, there's a flash of colour and I understand why Beth is doing what he says. The knife pressed close to her side is short in the handle but long in the blade, a gleaming steel mirror that catches the stained glass in my front door and throws kaleidoscope patterns on the walls. His hand is as steady as his voice as he says, 'There you go, Laura. No harm done.'

Beth speaks to me in a whisper even though he's closer to her than I am. 'I'm so sorry, I don't know what's going on, he's not supposed to be out, he—'

'Let's do this somewhere a bit more comfortable, shall we?' says Jamie affably, looking around my house like he's an estate agent sizing it up. 'You should be comfortable in your condition. Where's your kitchen table?'

Of course I can't help my eyes travelling towards the little flight of stairs. He uses the tip of the blade to steer Beth away from the front door and down the hallway. 'If you could just both make your way down there,' he says, and then, 'Your faces! Really, don't worry! It's just a bit of outstanding admin.' His tone is the one he used at first on Lizard Point: matey and authoritative. Everything is fine as long as it's under Jamie's control. But if he could control himself, he wouldn't be here. I have experienced how quickly Jamie Balcombe can flip between these two selves. The knowledge sharpens the knife.

I inch sideways along the wall, take the five steps down to the kitchen the same way, my phone behind my back, trying to guess on the virtual keypad where the nines might be. But I can't even unlock it with my PIN, and anyway, Jamie's on to me.

'Thank you,' he says, when all three of us are in the kitchen. He holds out his free palm like a primary school teacher confiscating some chewing gum. Instinctively my fingers grip the phone tighter. He prods the knife into Beth's side. At her stifled yell, I throw it down on the kitchen table. Jamie places his own phone next to it; from the same pocket he produces Beth's. He lines them all up on the Formica table, faces up, a row of glossy blank slabs. Mine looks old and battered next to Beth's sleek, oversized handset and Jamie's BlackBerry.

Does he mean to rape us? Does he mean to *kill* us?

I know this kitchen, I can find my way around it blindfolded. There is our block of knives, across the room and out of my reach. They are painfully visible but I also know that there's a cleaver in the drawer behind me, and a mallet in the pot with all the wooden spoons. Holding the door open is an old-fashioned ten-pound weight that would double up as a cosh, if only I could bend down to retrieve it. There are all the murder weapons you could want in here and yet Jamie has the advantage. All it takes is one blade.

The minute hand on the kitchen clock ticks forwards. Kit is due back any time now. This is scant comfort. I know that he would do anything to protect me. I also know, from Lizard Point, that this is not the kind of crisis he thrives in.

I don't plan to kick Jamie; rather, I see my foot rising towards his groin. My aim is true but he moves at the last moment, and rather than the incapacitating kick in the balls I was going for, most of the impact hits his thigh. The shock makes him drop the knife, which is good, but I wanted him on the ground and before the blade hits the tiles there's an explosion of pain in my head. The blow Jamie lands on my left ear is so forceful that the explosion seems to come from somewhere deep inside me. I stagger into the wall. The knife appears to slide across the floor again and again as my vision plays tricks. The room tilts gently to the side, righting itself just before the point of my collapse, then tilting again. When the images circling in my head join together as one, I see that Jamie's knife is on the far side of the room, the blade wedged under the fridge, and he has taken another knife from the block, twice the length of the weapon he came with. Kit's prized Sabatier knife. I bought it for him last year to celebrate fourteen years of marriage, the steel anniversary. You can cut through gristle with a Sabatier. We had it professionally sharpened a couple of weeks ago. You could probably perform surgery with it.

'What've you done to her?' asks Beth. The pain from my ear radiates through my neck and into my teeth. Her words are misshapen. Is this a perforated eardrum? My tongue probes a loosened molar but I can't taste blood, only sense a hot rhythmic throbbing.

'Sit down, please.' The new knife in Jamie's hand is so still it's as if it's been set in mid-air, cast in thick glass. As Beth slides behind the table, his other hand reaches into his back pocket and produces a few sheets of printer paper, which spring out of their folded quarters as they flutter to the table.

'You too, please, Laura,' he says. 'Beth, keep your hands under the table.'

'I can't fit any more,' I plead. Moving my jaw to speak plucks at the damaged nerve. I remember what he did to Antonia; this is his signature move.

He looks me up and down. 'I'm sure you can.' His downward tilting of the knife to my belly is all the persuasion I need. I say a silent sorry to my twins as I ease myself painfully on to the bench. There's a splash of yesterday's chicken soup I must have missed when I wiped the table down. Three seconds into this sitting position and my back begins to cramp. How long does he mean us to stay like this? There is nothing I can use to defend myself here. The confiscated phones, their screens black in repose, are out of reach.

'Hands under the table, Laura,' he says. Again, I can only obey. As I slide my hands on to my lap, the skin on my arms feels . . . fine. Normal. It's like I've shot straight through worry and out the other side. It makes an awful sense. The creeping feeling has always been about dreading the worst. Now the worst is here, it's too late for the early warning system. Even my anxiety has given up on me.

Face to face with Beth for the first time, I see with shock that her white neck is ringed with a vicious red welt. The imprint of a buckle is clear under the point of her chin. She sees me see it. 'He made me drive,' she says, and it's enough; I picture her, a belt looping her neck to the headrest. 'He was at Antonia's when I went over.' I retch to picture the scene: the Balcombes' comfortable family home broken into and terrorised by its former master. My voicemail echoing in a blood-sprayed room. Beth reads my face and shakes her head. 'They'd already gone,' she reassures me. The image downgrades to that same notional room, a back door swinging in the breeze, toys abandoned in the garden. 'She texted me to tell me he was there but by then it was too late.'

I can't get the timeline straight in my head.

'Never mind Antonia,' says Jamie. 'We've got to get down to business. Three pages each should be enough, for now.' He nods his permission for us to take our hands off our laps and we share out the paper between us. Whatever he has in mind, I resolve to spoil my paper; if we run out, one of us will have to go to the study to get more out of the printer. He can't overpower us both up two flights of stairs. I try to telegraph this thought to Beth but her eyes are wild and unfocused.

Beth's phone buzzes and the screen lights up.

'Hands under the table!' shouts Jamie. We both obey; I don't need my hands to read. Two messages are displayed on screen, one queuing behind the other. I read the earlier text first.

Where are you?
Jamie out 2 days early & AWOL already.

Probation messed up.
DO NOT GO TO MY HOUSE.

The most recent message has switched from warning to panic.

RANG YOUR FLAT.
SARA SAID YOU WERE COMING TO ME.
DO NOT GO!!!
FOR GOD'S SAKE BETH, CALL ME.

Jamie's eyes are fixed on the phone.

'Who's Sara?' I mouth to Beth, while he's distracted.

'Flatmate,' she mimes back. That's who answered her landline.

Ten seconds later, the screen flashes again.

I told police what Jamie's been saying about you, they're taking it seriously and going to make sure you're safe. Pls call so worried.

Jamie snatches the handset, throws it to the floor and crunches it underneath his heel, eyeballing us, daring us to say anything. There's a glittering seam of hope in the despair: the police will have a tracker on her phone, I think. They'll be on their way now. I've watched enough *Traffic Cops* to know that they can put an ANPR trace on her car; even if it's parked outside now, wedged into a tiny London space, her journey will have taken her past dozens of CCTV cameras and they will be able to narrow it down to North London. The car will have been ticketed anyway; they're very hot on that on the Ladder. That might throw up an alert. Antonia can tell them the rest, I think; they might be on their way already. Then my heart sinks, for Antonia will be telling the police to locate Laura Langrishe and Kit McCall. If they don't get the car, it could take them days.

Jamie trowels on the charm. 'Right, we'll follow the same batting order as in the trial,' he says. 'You first, Beth. How do you want to do this? You can write the truth down for the first time in your life, or you can take dictation.'

'I've *told* the truth,' she says angrily. Inside I'm screaming with frustration. Why not lie? There's no way anything she writes here can have any kind of sway in court. An essay about alien abduction would carry the same worth. 'I can't do it,' says Beth. Her eyes skid over my belly and bounce off. Whatever her fucked-up desire to hurt me in the past, she can't want to harm these children. Stupid bitch, why doesn't she just do what he wants?

The minute hand jumps to cover the hour: ten past two. There is nothing outside the house to warn Kit of the danger within. His shock will be greater than Jamie's; but Jamie is closer to the edge.

58

KIT
21 September 2000

The broken glass backfired. Rather than sever ties with Beth, Laura wanted to talk it through; to understand *why* she'd done it. I ought to have predicted it; she wanted to help her friend. I'd misjudged the whole thing massively; raised the stakes and leapt straight to a point where the girls could basically never talk again without working out they were both being lied to. That, of course, would mean the end of everything. The morning after the glass, I'd blocked Beth's number from Laura's phone, but there wasn't much I could do about the landline except ensure that Laura was never alone in the flat.

I was already halfway to mad at this point. It was over a month since I'd last had more than four hours' continuous sleep. My new hobby was to lie awake until Laura fell asleep, then fire up my laptop, open a blank document, the better to create a plan of action. It always began the same way:

Objective:
Get away from Beth without a confrontation.

Method:
Need Beth to do sth worse
Do midnight flit; stay with Ling? Mum?

If not look into loan so we can pay 2 lots of rent

How to sell this to Laura:
...

It was impossible. The screen stayed as blank as my mind.

Perhaps the most disturbing part of my new nocturnal routine was that in the absence of a plan, I'd often find myself writing a confession. It was never a conscious action; rather, I would come to and realise I'd filled half a page with the details of my night with Beth. I could never read it back; instead, I'd hold down the delete key and watch as the words were sucked back up into my hand. Then I would close the laptop and go back to bed, only to thrash awake a couple of hours later convinced that I had saved the document and given it an incriminating title. Knowing that Laura never went near my laptop wouldn't take the edge off my panic; I'd have to go back to the computer and trawl through every file, searching for something that wasn't there.

When Beth at last called, I hovered next to Laura, who put the telephone on speaker mode. That the appeal had been denied was the news we'd been longing for; we were free at last. This was our point of liberty; I wanted to grab the receiver, smash it against the wall, but I could only listen in horror as the girls' first conversation since the broken glass skidded headlong into confrontation and my inevitable unmasking.

'Can we go out to dinner to celebrate?' said Beth. If you knew the whole story, her confusion was obvious. 'My treat. To say thanks for everything you've done for me.'

Laura went quiet as my heart doubled its weight. I had to stop them talking. I flexed my foot to remind Laura what I'd gone through. When I brought it down, I trod on a loose wire and saw, quite spontaneously, how I could exploit this. The idea was like an EXIT sign flashing green in a dark cinema.

'It's just, dinner?' said Laura to Beth. 'After what happened last time. We hardly parted on the best of terms.'

I moved my foot an inch to the left, then hooked my big toe around what I was 99 per cent certain was the telephone wire. I could see even from here that the connection was loose at the socket.

'What do you want me to say?' Beth's confusion was fast hardening into irritation. I trod down hard, waiting for the line to go dead.

'Sorry would be a start,' said Laura, matching Beth's anger. Fuck. I tried to look at the wires without making it look like I was looking. I tried to move without making it look like I was moving. My pulse was so loud in my head I expected my neck and wrists to bulge with the beats.

'*Me* apologise to *you*?' said Beth. I moved my foot a little to the left and tried again.

'She's hung up on me!' said Laura. There was no time to enjoy the relief. Before she could check for a dialling tone, I took the phone out of her hand under the guise of soothing her and set it gently down.

'Maybe you should calm down a bit first,' I said. 'Your hands are shaking.'

I had a poker face, a poker voice and a poker body, from the eyebrows I strained to keep in place to the toes I willed not to tap. I knew that Laura would not be able to let this lie for more than a day or two. I mentally replayed Beth's mistakes – the gifts, the photographs, the constant fucking turning up at my house – and reasoned that all I was doing was building on foundations she'd dug herself. What she had started, I needed to finish tonight.

At three o'clock in the morning, London is as quiet as it gets. The world seemed soundless as I crouched on our third-floor landing with one of Mac's old lighters in one hand and a fifty-quid candle in the other. Beth had been in our flat on her own at least once; time enough to copy a key from our spare and set the original back on its hook. Suggesting this trespass to Laura would surely persuade her it was time to cut all ties.

The gas hit the flint and the flame glowed gold. I crept upstairs with a heavy heart. If I could have thought of a way of getting away from Beth without hurting anyone, I would have done it. I would have done it in a heartbeat. But I had not had a better idea. I had the wrong kind of intelligence for this sort of thing, and this was the best I could do. I took one last look over my shoulder at the little flame on the third-floor landing, and then turned the corner into darkness.

I hoped that the smell of Blood Roses might waft up the stairs to wake Laura up. If not, I'd leave it half an hour or so – long enough to make seem as though Beth had long been and gone – and then together Laura and I would go down the stairs and find the candle. I was almost looking forward to it. We'd blow it out, talk it through, and together we would conclude that Beth's campaign against us was escalating, and that we had to go. I thought we might stay awake, pack the important things, and be away before breakfast. I didn't want to *live* with my mum but I didn't mind spending a couple of nights there while I persuaded Laura never to muddy our clean break from Beth.

I was still so complacent I stayed a few seconds in my counter-feit sleep after Laura woke from her true one. I had not taken into account how deranged I was, in those sleepless weeks. I suppose that my temporary state of insanity drained the part of my brain that usually concerned itself with physics and chemistry. I'd been focused on the fact there was no draught in our stairwell, no soft furnishings to catch, and I'd overlooked the old-fashioned paint and peeling paper on the stairwell walls. An untrimmed wick burns high and the flames of a new candle double their reach. The heat alone must have been enough for that decades-old paint to bubble; the stairwell went up like it had been doused in petrol.

The acrid smell of both smoke and the burning paint filled the flat in what seemed like seconds and then it was too late; it was far too late. Laura was running away from the safety of the rooftops into the roaring heart of the fire. The injury to my foot had been deliberate but when I took the hit of the hot door handle, it was all instinct. The smoke and the fire and the destruction were so far from my intention that they barely seemed like my doing. If you'd asked me who had set the fire, in those burning moments – if I'd been able to think, if I had been able to talk – I would have told you that it was down to Beth, and believed it myself.

Fire changes everything. That night was the beginning of Laura losing her confidence; the start of her dependence on me. The paradox is uncomfortable that I am the cause of so much of her anxiety, but I have tried to see the good in it. I had to become her nurse as well as her protector and, while I hated to see her in pain, there's no denying that in our new dynamic we regained the intimacy she never really knew we'd lost.

I didn't see all that coming in the hour after our escape, as Laura and I sat in the ambulance and watched the windows of our flat exhale black plumes. Nothing rips you out of the future like pain; your world gets reduced to the searing moment. Instead of cooling, my hand felt hotter every second, as though it was still pressing down on hot metal; I would not have been surprised to see acid burn through the bandages. I tried to flex my fingers to see if the nerves were damaged, but even invisibly small movements tore at my broken skin. I could already feel the welts and ridges in my palm's flesh. I took only small comfort in my observation that, in the future, if ever Laura forgot how close Beth had come to destroying us, all I had to do was hold her hand.

59

KIT
21 March 2015

On the Piccadilly line train the wheels hurl a repeated accusation at the tracks: *Idiot, idiot, idiot*, they chant, in a fast waltzing rhythm that seems to grow louder with each strike. Years crash into each other and lies leap over one another. I still don't know how Beth found our house, or whether that's significant. I don't know what it is I must try to disprove. Five minutes from Laura and I still don't know what I'm going to say to her. I'm dreading the look on her face but at the same time I'm hoping that I step through the front door to a stream of rage, that she loses her advantage of surprise by spelling out the unknowns that make it impossible for me to establish a case for the defence. Does she only – only! – know I slept with Beth, or have they, between them, worked out the full extent of my deception? However Beth made contact, it would have been terrifying for Laura and although that will now have been supplanted by rage, it doesn't seem right. All those years of protection will have been for *nothing*.

The glass through the letterbox and the fire; these things they would have had to piece together between them, one awful truth propelling them along to the next. But it would have been a volcanic conversation, re-routed by tears and cross-purposes. It's possible they never got that far. Possible, but not probable.

Rising up the escalator at Turnpike Lane, I taste my neighbourhood, petrol and garlic, but this familiar air does not offer its usual welcome. Is there a chance I could style it out, as Juno would say? Laura is used to the idea of Beth as unreliable and dangerous; why shouldn't her story about sex with me be one more lie? It's not like Beth can prove it, all these years later. I haven't got a weird birthmark or anything they can use to identify me. Unless there's some other way they can compare notes, maybe something I do that other men don't, or something other men do that *I* don't? I feel sick thinking about the members of this tiny club discussing the thing only they have in common.

I touch out with my Oyster and on the beep realise something so obvious that it stops me in my tracks, and the jaws of the barrier trap my rucksack. It's a measure of how stressed I am that I don't even think about my camera equipment inside as I heave myself forward to wrestle the bag free. The crucial point I have overlooked is this: Beth has truth on her side. Beth has *always* had truth on her side.

I step out across Green Lanes, almost into the path of a red double-decker. The breaks squeal, the exhaust belches and I would almost welcome its impact. Ignoring the bus driver's shouts and the motorists' horns that blare in discord, I walk on, wondering how any of this ever seemed so deeply necessary. I think about the man I used to be, work spiralling out of control, a twin brother intent on his own destruction, all the other pressures that made my drastic solution seem like the only one.

On Duckett's Common teenage boys are playing basketball, the ball hitting the asphalt with that distinctive echoing bounce that's redolent of some Bronx recreation ground. In the playground toddlers play with dads who look a lot like me, and I take heart; my twins inside Laura feel like an insurance policy on my marriage. Those babies are half mine; they are half *me*. She won't want them growing up with only one parent, not after all she's been through. Our children are the reason we will get over it. But to get over it we have to go *through* it.

My pulse notches up with every single step I take. I turn into the corner of Wilbraham Road and stand on the opposite pavement, so that I can see, at a diagonal, our rooftop with its encroaching exoskeleton from next door's scaffolding. The skylight's open, telling me Laura's at home.

I make one last attempt to call her, to gauge what I'm walking into. The phone rings and rings, muffled and deep inside the house. Laura always has her phone with her and I've never seen her refuse to answer it when she knows the caller. I am in deeper shit than I have ever been before. With utter certainty I know that I can't face her. Or, not yet. Not without a little fortification.

In the Salisbury, I sit at a high table, order a neat double vodka and knock it back. It doesn't touch the sides, so I order a pint and hope that in the time it takes me to drink it, my palms will stop sweating, inspiration will strike, or, if I'm really lucky, a meteor will land on the pub and I'll never have to go home again.

60

LAURA
21 March 2015

'Look, I know this isn't easy for either of you.' Jamie's coaxing, reasonable tone is at fatal odds with the knife in his fist. 'Trust me, I don't want to go digging over old ground any more than you do. But people have believed your story for long enough. Don't you think? It's time my name was cleared.'

'Jamie, you *know* what happened,' says Beth. With that, Jamie loosens his grip on the shaft, just for a second; long enough for the knife to drop a millimetre downwards. It tears a two-inch slit in my T-shirt and makes a single thorn-prick of blood on my belly, at the highest point of my bump. I can't help the scream that comes out, but at Jamie's harsh, 'Shush!' I do manage to stifle it; still, I feel that scream like a moth in my mouth. *Write it*, I telegraph to Beth. Write whatever he tells you to, and the more extreme the better. Every lie he forces you to write is more ammunition for having him put away for life.

If we get out of this alive, says a little voice inside me.

The cut isn't deep; it's just a score on the surface. The blood and the position on the crown of my belly make it look worse than it feels. I'm far more worried about the blow to my head. My ears are still ringing.

The cloth of my T-shirt catches the droplet of blood, and as it's absorbed it appears to bloom, sudden and fast, like a time-lapse video of a poppy going from bud to full flower. It's this spreading stain that seems to decide Beth. She picks up the pen and squares off the paper before her, then looks levelly at Jamie.

'Why don't you tell me what really happened,' she says, in the deadest of voices. Her tone doesn't seem to register with Jamie, who smiles. When he clears his throat, I feel the vibration in the tip of the knife.

'On the tenth of August, 1999, I travelled *alone*,' he lingers on the word to wring meaning from it, 'to a music festival on Lizard Point, Cornwall, held to coincide with a total eclipse of the sun. There was a free and easy atmosphere at the festival.' He pauses again, not for effect this time, but to let Beth's pen catch up with his dictation. 'On the second evening, around a campfire, I struck up a conversation with Jamie Balcombe, who was also travelling alone. We hit it off immediately.' It's the voice he used in court, that expensive education undegraded by prison.

I don't know if Beth is aware that she's shaking her head as the lines of spiky writing fill the page. She's barely looking at the letters as she forms them. Her eyes dart in a triangle, from the paper in front of her, to Jamie, to the knife poised above my belly.

'Not so fast, you're not taking a secretarial exam.' His laughter comes from his neck, not his belly. 'It's got to be legible.' Beth carries on forming the letters at the same pace but slightly more evenly now. I can hear her teeth grinding in her jaw, a sickening bone-on-bone crunch.

The second hand on the clock jumps. Twenty past two. Kit will be here any minute.

Beth finishes her paragraph. Jamie continues. 'On the morning of the eclipse, I bumped into Jamie again. We decided that it would be nice if we went together to watch the eclipse somewhere a bit more private.'

Jamie's plan, once he's got Beth's 'confession' and mine, must be to take it to the authorities, in which case he will be giving himself up. This is the best-case scenario. The worst is that Antonia's right: he has nothing left to live for; he simply means to clear his name, as he sees it, on the way out, and if that means taking us with him, then I don't doubt he'd do it.

Everyone looks round as my phone plays Kit's tune. He must be just off the Tube. Each ring sends the phone scuttling across the table, a little farther away from me; we all watch its path towards the edge. I picture Kit innocently wondering if I need anything from Tesco on the way home. What if this is our last ever chance to talk? What if it's our last ever communication? The phone finally tips into the crack between the table and the wall, falling beyond reach and sight. It rings out a couple more times, then falls silent.

He will think my failure to pick up is me sulking about his social media fuck-up; how petty that all seems in the light of the knife over my belly. I would forgive him anything right now.

61

KIT
21 March 2015

The extra pint and the extra time do nothing to take the edge off. Panic is an antidote to alcohol. Halfway down my glass, a new terror bubbles to the surface. Maybe this isn't a sulk after all. Maybe Laura's in hospital; maybe there is something wrong with the pregnancy; maybe *because of what she has learned from Beth* there is something wrong with the babies – I have to put out my hand to steady myself. Spontaneous early labour due to shock seems like something from a Victorian novel, but these things can happen. *Anything* can happen, I ought to know that. Maybe she was rushed away before she could grab her phone. Christ, maybe she was *unconscious*. Why didn't I think of this earlier? Suddenly going home is as urgent as staying away was before.

I leave the dregs of my pint, hoist my rucksack on and walk to Wilbraham Road. Now, when I want to be fast, the air itself seems against me, tugging at my ankles. Someone has parallel-parked a white Fiat very badly outside the house, and there's a parking ticket under the wipers. On the doorstep, I shrug off my rucksack, feeling like I might float off without it. A cement mixer churns in next-door's front garden, the builders out of sight.

When I put my key in the door, I find it on the chain. This means either that she wants to keep me out, or that she's had the door on the chain all morning, and she's lying somewhere, unable to move, or worse. I squat to the letterbox; there's a movement in the kitchen, a displacement of light by shadow. Relief that she is up and about is replaced by the old fear: she's furious.

'Laura?' My voice echoes into the silent hallway. I see only a figure swaying in the kitchen. I slide down the side of the door frame and talk through the gap. 'Darling, please open the door. Let's talk about this.'

There's a gargled sob from the kitchen that screws my insides tight. I would rather anger than tears. I take the Swiss Army knife out of my pocket and flick out several blades before deciding on the short hooked point that doubles as a tin-opener.

'You have no idea how sorry I am,' I say, fiddling blindly with the chain on the door. 'It was a one-off, I promise. I've only ever loved you, you know that.' My knife gets purchase on the chain; with the hooked tip, I ease out the weight at the end. 'I wish it every day,' I say. 'I wish I could turn back the clock and not sleep with her. It only made me realise how much I love you. Please let's not let it spoil things, not when it was so long ago. We've got so much to—'

The chain gives so suddenly that I almost fall through the front door as it swings open. I set my rucksack on the floor and right the light-up globe that's toppled over. I actually brace myself before going down into the kitchen, arms flexed like a wrestler showing off his muscles, as though the onslaught will be physical. I am prepared for little fists and slim fingers clawing at my face.

What I find is worse, and so utterly unexpected that it takes a few seconds for my brain to process the image, let alone the meaning behind it.

Laura sits at the table. My eyes ricochet between her belly and her face; a thin oval bloodstain beneath a slash in her T-shirt, and her left cheek purple and swollen. With her are Beth and – recognition is instant, and terrifying – Jamie Balcombe. In a fancy shirt and chinos, his meaty fist clutching a knife tipped with blood.

'Kit, no,' says Beth, shaking her head.

What, *what?* How the hell have things progressed from Beth finding Laura to this? Beth's got some kind of graze around her neck and Laura's cheek is blackening down one side. I had always pictured Laura's face, when she found out about Beth, to be blown wide with rage but instead it's crumpled, falling in on itself. Her eyes are dry.

There's a piece of writing paper in front of Beth. Angular blue letters run in sloping lines. My past and the bewildering present don't connect, but bounce apart like repelling magnets.

'What the hell is happening here?' I only address Jamie because he is so clearly the one in control. No one answers. 'What the hell is *happening* here?'

Laura looks from me to Beth and back again and turns her head slowly away. It's worse than any outburst.

Jamie takes over.

'Kit!' he says brightly, as if it's his house and I an expected visitor. His demeanour is exactly as it was in Cornwall, in the moments after Laura interrupted him. For a second I see him as he was then; I picture the jeans he was wearing, his shoes, the spikes in his hair. The remembered image is so vivid it threatens to overwrite the scene in front of me.

My phone is in my rucksack, which is just inside the front door. I turn to retrieve it. 'I need to call the police.'

'No you don't.' Jamie's tone is dismissive even as he inches the blade closer to Laura's belly. It's broken skin, that's all; the stain isn't spreading. I'd need the reflexes of a spider and the strength of a bear to wrestle it from his hand before he could do any damage. I'm too far away and I'm too weak. 'What I mean is,' continues Jamie, 'there's really no rush. The police will get involved with this sooner or later, don't worry about that. Once my lawyers have been over this. We're halfway through already.'

The girls won't look at me, or at each other. Occasionally, Laura lifts her eyes halfway to Beth, then casts them down again, like her pupils are lead weights. Clueless and only now understanding what fear feels like, I have no choice but to play along. 'What are you *doing*, Jamie?' I hope I've matched his genial tone.

'Just getting the girls to set the record straight, say what they should've done back in Cornwall. Admit their collusion. Admit that the whole fucking mess was a little story they cooked up, some weird female idea of a practical joke.'

Fifteen years, he's been holding on to this. Behind the campaigning and the talks and the website, this has remained his Plan A. Even in terror, there's room to be impressed by this mineral patience. He jerks his head towards Beth. 'This one's been pouring poison in my wife's ear for months. They were a bit reluctant, didn't want to take the rap for what they've done, scared of the consequences. Perjury in a rape trial, it's a biggie. And not just any old rape trial. They'll want to make an example of you, Laura.' She flinches at the nonsensical accusation, but even blunt words hit home at knifepoint.

The blade in Jamie's hand starts to quiver, as does his voice. His mask is slipping in slow motion. 'So you can see why they're worried. But it'll be ok for you. It's not like you'll be banged up on a sex offender's wing like I was.'

Laura gives a full-body shudder that I feel through my skin.

'But Jamie, there's no way this will be admissible in any court. These –' I have to stop myself throwing air quotes around the words – '*statements* won't be worth the paper they're written on. You must understand that.'

'This is just the *first step*,' he says. 'You'd be amazed at what a good lawyer can do.' His lawyers were of limited good back in the day, I remember. I might not have full grasp of the situation, but I know better than to challenge him. 'Look,' he continues, 'I don't want to hurt the girls. I'd never hurt a woman. But I do need them both to finish their statements retracting what was said in court. This is my life we're talking about, here. This is my *reputation*!'

He's close to losing it. 'Ok.' I don't know what I'm doing apart from buying Laura and my children time. 'Why don't you put the knife down and let the girls go. I'm sure we can sort something out without them.'

'Actually no, I've been waiting fifteen *fucking* years to clear my name.' The swearword throws Jamie's voice an octave up the scale, making him sound both terrifying and ridiculous. He pauses for a second to breathe and gather himself. 'So you'll forgive me if I just see this through to the end. Come on, Beth, keep going. Sooner you finish your statement, the sooner we can all get back to normal. Or rather, *I* can.' Beth picks up the pen again, then poises it mid-air as Jamie holds up his hand to continue speaking. 'Actually, Kit, I'm wondering if I need you to redo yours. Although, to be honest, theirs will cast yours in all the new light we need. You were never really in on it, were you? All you did was back up what your missus had to say. You never saw us at it.' At the table, Laura flinches. 'Although I wish you had, I'm sure you would've seen what was what in seconds.' At last, Laura looks my way. Her eyes fix on mine, but she's crying properly now and the look could mean, please save me, or we'll get through this, or I hate you. All the power in the room is concentrated in a length of sharp steel. With Jamie so much closer to Laura than I am, there might as well be a barbed-wire fence holding me back.

'Beth. If we could just finish up, please,' he says. Is it my imagination or is his voice losing its previous conviction? 'If you could just remind me of the last sentence you wrote and we can carry on from there.'

The pen skips in her hand but she doesn't say anything. She doesn't even look up from the page.

'Fine, I'll dictate it, then.' He clears his throat. 'I am profoundly sorry that I took something perfectly natural and enjoyable and twisted it around. I apologise to Jamie and Antonia Balcombe and their extended family for the distress caused by my false allegation. I am willing to make a statement in court, and –' the tip of the knife traces tiny patterns, as though Jamie were carving his words in the air – 'face up to any criminal or civil actions that come my way as a result of this retraction of my original statement.' He turns his head to me. 'To be honest, we can end Laura's statement the same way.'

Beth's hand has stopped moving.

'Write it *down*,' says Jamie. The knife in his hand trembles dangerously. Laura's tears glisten in the notch at the base of her neck as Beth sets down the pen.

'Please do what he says,' Laura urges.

'Laura's right.' It's the first time I've spoken directly to Beth since the day I broke the glass. 'It won't hold any legal sway. You won't be in trouble in court.' I come close to saying, do it to save your lives, but some instinct in me knows that to verbalise what is so obvious would be a mistake; it would pop the fragile bubble that separates reality from denial.

'I can't,' says Beth. 'I can't lie.' Something in her has become untethered, even since I got here. 'You raped me,' she says, and the simple three-word fact of it stops everything. It swells to fill our little kitchen. The only sound is the cement mixer churning in next-door's front garden. 'You followed me somewhere quiet, you held me down and you raped me. You did it again in court and you did it again on the internet and you've been fucking me over ever since. You've done it to your wife and you've done it to that girl and you've done it to God knows how many other people.'

'Beth, *please* just do what he wants,' says Laura. The tiny part of me not consumed with survival wonders about the girl, and the wife. But Beth only has eyes for Jamie. Suddenly, it hits me what she's doing. She's burning up everything she's got left, like a rocket on re-entry; she doesn't need it where she's going.

'You *raped* me.' The words come out in sharp jags.

It seems that I alone notice that Jamie's knife is travelling, unsteadily but surely, like a compass needle, away from Laura and towards Beth.

'I couldn't write what you want even if I knew you were going to set this paper on fire.'

I calculate fast; if I grab his upper arm rather than his hand, I can force the knife away. There are three of us and only one of him. But I have always been able to think faster than I move, and I'm still a stride away when Jamie pulls his arm back and jabs the knife into Beth's side. It bounces off in a way that suggests he has hit a rib, but he lunges again and this time the top two inches of the blade disappear.

I don't know who the scream comes from.

He withdraws the knife; blood stains the steel. Beth is a dead weight, hitting the floor.

I am fast but my wife is faster. Laura gets there before me; she knocks, rather than wrestles, my Sabatier from Jamie's hands. It somersaults in the air; the dull shaft up, the shiny blade down, shaft down, blade up, and for a sickening moment it looks like Laura's going to catch the cutting edge in her open palms. But she only brushes the end of the handle with her fingertips.

'You bitch!' Jamie is already halfway to the knife block. Laura's face is blank with panic as it bounces from her grip to land on the table, where only I can reach it.

The knife is both familiar and uncanny in my hand as I charge across the kitchen and thrust its sharpened tip into his throat. There is a split second of resistance, which I guess is the knot of his Adam's apple, and the blade slips sideways, missing the spine; the rest is like cutting ice cream in comparison, and a second later, the tip is a shark's fin protruding from the nape of his neck. I withdraw with futile haste; it is done. The knife clatters from my hand to the floor at the same moment Jamie hits the tiles with a thud. His body lies next to Beth's. You can't tell whose blood is whose. Everywhere is red, the kitchen floor a glossy sea. Jamie gargles, then vomits a crimson geyser that coats everything – me, Laura, the walls, the furniture – in a fine pink spray. I watch, transfixed, as his blue eyes turn to marbles.

And then I freeze.

Laura steps over Jamie's body and crouches in the puddle of blood.

'Beth?'

'Laura, I—'

'Call a fucking ambulance!' she screams up at me. And only now do I manage, reaching for the phone that's closest, a BlackBerry, and punching in the numbers, and telling them that we need an ambulance because there's been a double stabbing, two casualties. I give my address and I've even remembered the phonetic alphabet and I ask them, foolishly, if they'll need a parking permit and they reassure me that they won't.

While I say all this, Laura is on her hands and knees, cradling Beth. With calmness and foresight, she has taken the spare tea-towels from the drawer and folded them into little pads to try to stem the bleeding. The first is already saturated.

'They're on their way,' I tell her. 'How's she doing?'

'I don't know, I don't fucking know,' she says, and then to Beth, 'Keep your eyes open, for fuck's sake, keep awake.' With bloodied fingers she pushes a damp wad of hair away from Beth's face. Beth's breath is coming in short sharp bursts now. She's trying to say something and she's looking at me. 'I didn't—' she says.

'Don't talk,' Laura's saying. 'It's ok. The paramedics are coming. It's ok. We've got you.' I can't read how Laura feels about Beth but there's no mistaking the look she gives me now: *I hate you. I hate you. I hate you.* 'It's definitely slowing down. We just need to keep her warm. Take off your coat.'

To do this, I must put down the knife: *Exhibit A.*

I pull my arms out of my sleeves and lay the heavy coat down over Beth, as gently as I can. It's impossible to tell how much blood she's losing; her clothes are sopping. My coat is smeared with mud from the mountainside at Tórshavn. I tuck her in and send her a silent apology for everything that has happened as her lips turn white.

A battering on the door makes me and Laura flinch but Beth is immobile. When I open the door to the police and paramedics, my hand makes a bloodied smear on the brass knob. Outside, the emergency services have blocked Wilbraham Road. A patrol car is diagonal across the road markings. There are two ambulances either side of it, blue lights rotating in silence. One of them is now merely a hearse. I usher the paramedics in their green boiler suits downstairs to the kitchen where there might be one life left for the saving.

Laura finally peels apart from Beth, standing over her while the experts take charge. She looks like she's wearing red evening gloves that reach up to the elbow.

'She's gone into shock!' says one of them. Laura claps one hand over her mouth and begins to babble under her breath.

A third paramedic enters the kitchen and looks in alarm at Laura's swollen face.

'Are you Christopher Smith?' the police constable asks me. Thickset, with rugby shoulders, he looks like the kind of boy who would have bullied me at school.

'You don't need to arrest me,' I say. 'I'll come with you.'

He surveys the bloodbath below.

'Oh, I think I do,' he says. 'Christopher Smith, I am arresting you on suspicion of murder. You do not have to say anything but it may harm your defence . . .' I let his caution roll over me as I offer up my wrists. All my attention is on Laura, standing at the bottom of the stairs. The carmine imprint of her own hand is across her mouth; her arms cradle her belly. The look in her eyes hollows me out.

'*When?*' she asks.

When what? It takes a few seconds for me to understand what she means. That cry through the letterbox. It was Beth, warning me to shut up.

Beth didn't tell Laura what happened on Lizard Point.

I did.

'In Cornwall,' I say it at last. 'The night before you got there.'

Laura nods once, then closes her eyes as if the sight of me is more than she can bear. The handcuffs' click reverberates through our red kitchen. I drop to my knees. My bound hands are dead weights in my lap, my wedding ring bloodied. Laura turns her back on me. As my heart comes undone, somewhere in the loss and the horror is the overdue liberty of relief.

Fourth Contact

62

LAURA
30 September 2015

We stand side by side in front of the speckled mirror. Our reflections avoid eye contact. Like me, she's wearing black and like mine, her clothes have clearly been chosen with care and respect. Neither of us is on trial, or not officially, but we both know that, in cases like this, it's always the woman who is judged.

The cubicles behind us are empty, the doors ajar. This counts for privacy in court. The witness box is not the only place where you need to watch every word.

I clear my throat and the sound bounces off the tiled walls, which replicate the perfect acoustics of the lobby in miniature. Everything echoes here. The corridors ring with the institutional clatter of doors opening and closing, case files too heavy to carry wheeled around on squeaking trolleys. High ceilings catch your words and throw them back in different shapes.

Court, with its sweeping spaces and oversized rooms, plays tricks of scale. It's deliberate, designed to remind you of your own insignificance in relation to the might of the criminal justice machine, to dampen down the dangerous, glowing power of the sworn spoken word.

Time and money are distorted, too. Justice swallows gold; to secure a man's liberty costs of tens of thousands of pounds. In the public gallery, Sally Balcombe wears jewellery worth the price of a small London flat. Even the leather on the judge's chair stinks of money. You can almost smell it from here.

But the toilets, as everywhere, are great levellers. Here in the ladies' lavatory the flush is still broken and the dispenser has still run out of soap, and the locks on the doors still don't work properly. Inefficient cisterns dribble noisily, making discreet speech impossible. If I wanted to say anything, I'd have to shout.

In the mirror, I look her up and down. Her shift dress hides her curves. I've got my hair, the bright long hair that was the first thing Kit loved about me, the hair that he said he could see in the dark, pulled into a schoolmarm's bun at the nape of my neck. We both look . . . demure, I suppose is the word, although no one has ever described me that way before. We are unrecognisable as the girls from the festival: the girls who painted our bodies and faces gold to whirl and howl under the moon. Those girls are gone, both dead in their different ways.

A heavy door slams outside, making us both jump. She's as nervous as I am, I realise. At last our reflections lock eyes, each silently asking the other the questions too big – too dangerous – to voice.

How did it come to this?

How did we get here?

How will it end?

63

LAURA
28 September 2015

Tonight there will be a blood moon; not a solar eclipse but its inverse, the earth passing between the moon and the sun. At three o'clock in the morning, the light passing through the earth's atmosphere will turn the moon a rusty red. There are no windows in Kit's cell in Belmarsh. He will only be able to observe the lunar eclipse if the twelve men and women currently deliberating his case find him innocent.

'I feel sick,' I say to Ling.

For once she doesn't try to reassure me; she's all out of platitudes. 'So do I,' she says.

I run my wrists under the cold tap. We are in the women's toilets at the Central Criminal Court, better known as the Old Bailey to the public and simply the Bailey to those in the legal profession – and I am beginning to feel like a professional witness.

We are high up in every sense; Court Twelve is on the topmost floor, up eighty-nine steps. There is no milling about in the vast lobby for spectators at the Bailey, no spying on journalists or accidentally bumping into the players. Here, the public galleries – to which I am relegated, now that my performance is over – are segregated from the main business of the court. No matter who you are, no matter how involved you are in the case being tried, there are no privileges in the public gallery. It's much stricter than the crown court at Truro. You're not allowed to bring even a bottle of water in with you. My mouth is so dry that if I try to lick my teeth, my tongue sticks. I put my head under the tap and drink greedily.

'Oh, Laura,' says Ling. 'God knows what's in that water.'

She's right; it tastes like dirty pennies. I gulp it down.

'How long have the jury been out now?'

She looks at her watch like it hasn't been thirty seconds since I last asked. 'Three hours.'

'They might not finish today after all. I should ring home.'

That's not as easy as it sounds. At the Bailey, you can't just pull a phone out of your pocket. They don't even have somewhere you can check it in: I've had to leave mine in a café over the road. Ringing home means going down three flights of stairs, through security, across the road, into the café, then, after the call is made, doing the whole thing in reverse.

'But what if they announce it when you're gone?' says Ling. 'Like when you're waiting for a bus and you go and get a drink from the shop and it goes past.'

It's almost worth it to tempt fate, but I have to be there for Kit.

'You're right,' I say. Besides, I don't want to run the gamut of all those journalists. This trial has been front-page news, trending on Twitter, national television headlines, radio-debate fodder. I have been offered tens of thousands of pounds to tell my side of the story. There have been press photographers outside my house for ten days now.

The twins are five months old and at home. I had no choice but to go back to work when they were ten weeks old. And yes, I do resent it. With a husband on remand in Belmarsh and a six-figure legal bill on the horizon, I took the highest-paid job I could get, head of a British alumni fund for an American university, skimming cash from rich graduates to upgrade an already enviable campus. Not the most deserving of causes, but a well-paid one, and a prestigious gig; my appointment made the news pages of *The Fundraiser*. For the first time in my career, I had to have a professional photograph taken. I wore my hair down. There is no longer any reason for me to hide, nor indeed anyone to hide from. It is a measure of my exhaustion that this past week of compassionate leave, with its gruelling days in the witness box and the ensuing grind of watching from the public gallery, has been respite from the slog of working motherhood.

When we come out on to the landing, Mac's there, smelling like fresh cigarettes. He's started going to NA meetings again as a pre-emptive strike. Adele is next to her son, dressed in black. We exchange tight smiles. There are two courts on each stifling landing, supervised by unsmiling security guards. Their grid of CCTV screens forms a compound eye of legal dramas, all playing out in bluish monochrome.

Everything at the Bailey is in a different league. Instead of two regional barristers, this trial is being fought by Queen's Counsels. Our silk is called Danny Hannah. He's in his late fifties and I think I have just paid for his children to go to university. His horse-hair wig is reassuringly disgusting and he has been insistent to the point of blitheness that Kit's acquittal is a foregone conclusion. One newspaper columnist expressed astonishment that the case should come to trial in the first place although, as Danny explained, the outcry if the CPS had failed to prosecute would have been far louder.

That Kit killed Jamie Balcombe has never been in doubt, but for a murder charge to stick, motivation has to be proved. The echoes of that other trial are at times deafening.

Kit performed well in the witness box. He, and I, and the other witnesses, only served to underline that he acted in self-defence and in defence of his wife and unborn children. The past has been raked over, of course it has, and while we have both explained all the details of that night in my kitchen, we agreed, in whispers across a prison table, the things that should stay between us. Some truths cannot be swallowed whole. Forensics are on our side this time; Jamie's prints on our knife and the bloodied slash in my T-shirt speak volumes. That the discarded knife on our kitchen floor came from Jamie's former family home shows premeditation only on his part. Beth's little dossier, laid out on my desk, nearly undid us; the CPS seized on it as proof of premeditation, hence the murder charge rather than manslaughter. In my cross-examination they suggested that I'd put it there on purpose, laid it out as the perfect excuse. I stood my ground like I was bolted to the floor. And of course we had Antonia to corroborate Beth's written word. But juries are strange beasts. You never know.

'All parties in Smith to Court Twelve.' The public address system has the crackle of old vinyl.

My hands are loose at my sides; Adele grabs one, Ling the other, and Mac holds onto Ling. We stay that way, as if in group prayer, until the security guard motions us in.

Not for the Bailey the intimacy of Truro Crown Court. This public gallery is all brass and marble, a balcony twenty feet above the players. The press bench is busier than it has been since the trial began. Our old friend Alison Larch is in the throng, a weird version of her past self. She's done something to her face so that her top lip juts like a bill; light bounces off her smooth brow. Mac holds on to the brass railing in front of him. The veins bulging on the backs of his hands threaten to burst through his skin.

The Balcombes are already seated. I think they may have remained in the gallery since the jury were sent out. It is not the full entourage this time but just Lord Jim, as he now is, and Lady Sally. The brother was there for the beginning of the case, but was ejected after calling Kit a murdering cunt shortly after the indictment. I don't know what's happened to the sister. Sally Balcombe shakes and walks with a stick now. She has not been here every day, but Jim has been a constant presence. At first, he took the front row, like us. But, as the trial progressed, and his son's reputation was slowly torn to strips, as the layers of delusion and violence were exposed, Jim retreated. When the intern from Jamie's company came forward to describe her ordeal, he shifted into the middle row and now he is tucked away in a back corner, beyond the sightlines of the press and, mercifully, of Kit.

I can sense Jim's frustration even from here. It will be killing him that he couldn't select the prosecution barrister, that such a basic thing as getting his son's killer put away is beyond the reach of his wallet. I don't want to feel sorry for Jim and Sally Balcombe but I can't help it. They must be doubly grieving as their memories are overwritten by the picture painted in this court.

My husband is in the dock below me, wearing the navy suit I bought him before I realised how much weight he'd lost. He is handcuffed to an officer: another one guards the door, which makes me want to laugh – I mean, as *if*. I lean over the marble sill so that he can see me. The dock in Court Twelve is shallow and wide, with eight flip-up chairs in a row. I'm pretty sure the jury can't see the empty chairs on either side of him, although from my viewpoint up here in the gods, they have always seemed to underline his ultimate innocence. This is how bad it could be, those empty seats suggest. He could be a conspirator, on trial with someone else, one of half a dozen men accused of the same crime. He could be a *real* criminal.

Those empty seats also remind me how close I came to standing there beside him.

Kit turns and looks straight up at me, his eyes heavy with our history and a guilt that has nothing to do with the dead man.

'All rise for the judge,' says the clerk, and even Sally Balcombe gets to her feet.

The judge – a Lord Justice, no less – steeples his fingers and orders the jury returned. The twelve file in, brisk with their own importance. I keep my eyes on Madam Foreperson, a tweedy brunette who has remained inscrutable throughout what I have seen of the trial, and who stays that way now.

'In the matter of Crown versus Smith, how do you find the accused?' says the judge.

Anxiety sends fire ants marching all over my skin and I claw my fingers to scratch. Madam Foreperson clears her throat. 'Not guilty.'

There is a dry groan from the Balcombes behind us. Mac rather theatrically touches his head down on to the brass bar in front of him, and so misses the moment that Kit looks our way; you couldn't call the expression on his face a smile, more a release of the grimace he's been holding since his arrest.

'Oh, thank fuck,' says Ling. 'Thank fuck for that.' She turns to me. 'Are you ok?'

How am I supposed to answer that?

He doesn't just get to walk free. There's paperwork to be processed first, down in the cells. Documents have to be faxed to Belmarsh and then back again. It will be half an hour at least before I can see him again and when I do it will be in public. You'd think that they'd have a special reunion room for this purpose. After all, the Bailey is riddled with rooms, huge echoing atriums and poky little cubby holes. I saw dozens of them lying empty when I was waiting to be called as a witness. But no. When all the forms have been filled, he will be released into the street. This is undignified in itself and doubly so when you consider the bank of cameras waiting for us on the pavement. It's a final humiliation.

But he is not out yet, and so we are free to sneak out to the café the long way round, through Warwick Passage, a brick-tiled subway next to the public entrance. There, I fall upon my phone as though it were one of my children rather than a mere link to them; I learn that the babies are fine, sleeping off their lunchtime feed. I feel a degree of release. After coffee, the four of us trail back to the Bailey to wait for Kit. The street outside the entrance is entirely nondescript yet instantly recognisable from the TV news, or maybe it's only the journalists who make it so. We hide from them in Warwick Passage, waiting for Danny Hannah QC to ring my mobile. I count the hairline cracks in the white tiles that line the walls. When my phone goes, Kit's already on the way down. It's a sixty-second warning. My stomach executes a series of perfect back-flips as I step into the street.

'Laura!' 'Over here, Laura!' 'Give us a smile, Laura!'

The flashbulbs hit me like a shove to the chest. I stand frozen. Across the road, a couple of Chinese tourists in socks and shorts take pictures of them taking pictures of me. Then the lights ripple as one, an electric-eel slither away from me. I'm still blinking in shock when Kit comes out and it's Mac who gets to him first. He rushes up to Kit and embraces him. No brofists or manly back-slaps for these two: Mac holds Kit like a baby, rubbing his back. It doesn't seem to bother either of them how public this is. One of the paparazzi actually tuts, and looks at me, his thoughts as clear as if he'd spoken them. *I'm* the picture he wants. I'm his money shot.

'Come on, Laura,' he says, and the attention is back on me. Kit pulls away from his brother and the individual flashes become a white wall of light as I take a single controlled step into his wide arms. I turn my face away – they're not getting this part of me – and hold my husband properly for the first time since the morn-ing he sailed for the Faroes. We dock together in the old way, the first way, because he feels as lean as he was when I met him. Kit's fingers find my face. He tilts my chin, brushes a stray wisp from my cheek and, closing his eyes, brings his mouth down on mine. We hold the kiss; dry but not chaste. We're too well-established for proper desire these days but some kind of muscle memory stirs in me. It's still there. It will always be there.

We move apart slowly; lips first, then bodies, and hold hands and pose for the cameras.

'Give us a proper smile,' says one of the photographers. We're all teeth and shiny eyes.

'Have you got a statement, Kit?' says a woman with a microphone.

Danny Hannah steps in front of us. 'I'll be speaking for my client.'

'What happens now?' says Adele, looking more like a little old lady than I have ever seen her.

'We get the fuck out of Dodge,' says Mac. He nods to the opposite pavement to where Ling has managed to hail a black cab.

We leave Danny Hannah to address the press, and pile into the taxi. A photographer on a bike scoots alongside us as we chunter with frustrating slowness down Old Bailey and then left down Ludgate Hill. The cab driver gives him the slip on Cheapside, indicating left then swinging the cab right, and by the time we reach Holborn we are alone. I check the street; a row of yellow lights shines behind me at head height. There are plenty of taxis. I tap on the driver's windscreen. 'Can you pull over here, please?'

Ling knows, but Adele, Mac and Kit exchange startled glances.

'Laura?' says Adele. It's her I feel worst for.

'I'm sorry, Adele,' I say, stepping backwards on to the pavement. 'I couldn't humiliate him in front of the cameras.' While she digests what this means, I turn to Kit. I told him this was coming but I can see from his expression that he has never stopped hoping I would change my mind. He can hope for the rest of his life. 'They'll be asleep for a while now. Just give me a couple of hours to get my head around everything. If you come round at six, you can be there for bath time.'

It's an instruction. He nods.

I slam the door on them and let Ling hail us a new taxi. It's not until we are safe in the back seat that I finally let myself cry.

64

LAURA
30 September 2015

Kit told me everything, across a scratched plastic table in a prison visiting room. There were ugly, gulping tears from him but the verse/chorus/verse of his confessions was punctuated with the same refrain. 'I did it for *you*,' he kept saying. 'I did it *for* you. One little fuck-up, that's all it should have been. It wasn't worth throwing away the best thing that's ever happened to either of us, for one moment of madness. I did it all to *save* us.' Of all his various excuses – Mac, work, his dad – I did it for you is the most offensive. How could anyone who loved me, anyone who *knew* me, do something so cruel in my name? It's a weird, stunted love that would rather put me through fifteen years of crippling anxiety than hurt me once. Too weak to live his life without me, he would rather live with me broken and ill than lose me. He has not only taken away my future with him but ripped out my past, too.

With visiting hours limited to two ninety-minute sessions a week, Kit had to drop his bombshells in instalments, like a Victorian novelist teasing his reader. Each chapter peeled off another layer of my skin. The afternoon I made him tell me the details – the tea tent, the sleeping bags – I made a connection: the soapy smell of him the day I arrived in Cornwall. He'd been in the shower, washing her off him. I thought then that *that* was the day I was peeled back to the bone. I cried so hard that evening that I graduated from tissues to kitchen rolls to towels. I thought that we needed a new word for this kind of crying; nothing in the dictionary came close to capturing the force of it. Maybe other languages have words for these tears of grief and anger and betrayal that feel powerful enough to kill you, but there is nothing for it in British culture.

He has said more than once that he wishes he had turned the knife on himself and in my worst moments I agree with him. 'You have no idea what I've been through,' he said last time in a desperate bid for sympathy and for the first time his words plucked at my sympathy reflex because he is wrong about that. I understand how it started. I obviously understand what it is to hold back. For years I thought it was *my* lie that could wreck us. I concede the parallel; but really, it is the difference between a valley dividing mountains and a line drawn in mud with a stick.

Would I have forgiven Kit for sleeping with another woman? It's a question I've asked myself endlessly. From the vantage point of encroaching middle age, I have compassion for his youthful insecurity (and a kind of dismayed wonder that I never picked up on it at the time). Ironically it's the sort of betrayal I might have forgiven later in our marriage, when I knew what time and pressure can do to a relationship. But at twenty-one I was passionately, inflexibly in love. He's probably right; it would have been the end of us. I wish it had. I wish he had merely broken my heart when he had the chance. Young hearts are like young bones; they bend. They mend.

I won't ever be able to forgive what followed; the systematic denigration of a vulnerable woman, of a *rape* victim, mortified about what she had unwittingly done to me, and traumatised by what had been done to her. Beth wanted to be a friend as well as to have one; he did not deserve her loyalty and though I wish she had told me the truth at the time, I can't blame her for keeping quiet. To think that Kit even made me doubt her original story about the rape; it sickens me. Worse, though, than the infidelity or the process that followed it, is this. Kit waited until we knew that Beth was going to live before he told me about the glass and the fire. For the first two weeks, her life hung in the balance and I visited him twice in this time. His remorse then was only for sleeping with her. If Jamie Balcombe's knife had been as accurate as Kit's, if its blade had gone deeper into her breast, the extent of his deception would have gone to the grave with her. The betrayal flayed my skin, but his cowardice chips the bone. It poisons the marrow.

65

LAURA
3 April 2015

I bought grapes, and only realised when I was in the lift that I
hadn't got the seedless kind. That's what you get for going food
shopping on Green Lanes. She was in the North Middlesex
Hospital, two floors above the maternity unit. For the first fort-
night, when she was in intensive care, fighting for her life as my
own superficial bruises healed, visiting was strictly family only. I
found out later that this wasn't hospital policy but something her
parents had insisted on. They didn't want anyone – me, Antonia,
certainly no one else associated with the Balcombes – to see her.
But she was three days out of the ICU now and security was
relaxed.

Still, there was a sense of trespass as I walked the long grey
corridor to her ward, clutching my paper bag of fruit. As I
approached, a dumpy little woman with dark grey hair shuffled
out. It took me a few seconds to laser away the puffs and pouches
of age and understand that this was Beth's mother, last seen in the
lobby at Truro. She had crossed the watershed of middle age into
elderly; her face and body had gone shapeless, her hair was white
and only the livid inner rims of her eyes gave her face colour.
Shame forced me behind a wheeled screen. I clutched my paper
bag so hard that it tore. A couple of grapes detached from their
stalks and rolled unevenly across the corridor into Mrs Taylor's

path. I cringed, bracing myself for the telling-off I deserved, but she was so lost in her own thoughts that she didn't notice the obstacles. I waited until she was gone, binned the fallen grapes and turned into Beth's bay.

She was sitting up in bed, eyes on the door. I could have sworn she was expecting me.

'Hello.' I set the grapes down on her bedside table, next to the hundreds already there.

'Gonna start my own vineyard,' Beth said, with a tentative smile; just as she had all those years ago, testing the water to see if it was ok to make a joke; now, as then, I found that it was.

'That hospital gown does a lot for you,' I said. 'Really brings out your eyes.'

She laughed, then winced, putting a hand to the lumpen dressings on her side.

'You're *huge*,' she said in awe.

'Tell me about it. I feel like I'm going to run out of skin.' I heaved myself into the bedside chair, feeling it sigh beneath my weight. Taking her hand in mine seemed right. Her skin was as soft as I remembered it. Now that I was here, I realised I hadn't rehearsed what to say. I ought to have arranged things into some kind of chronology in my head.

'I literally don't know where to start,' I said. 'Kit's told me everything. Or, I think he has; every time I think he's finished, something else comes out. I don't know what you know and what you don't.'

'I didn't know about *you*.' It came out with the urgency of something that had been on the tip of her tongue for fifteen years. 'I was mortified when I found out he had a girlfriend, I was heartbroken, I felt sick; I would never have slept with him if I'd known.'

Then, as an afterthought, less urgently; 'I wasn't after Kit, you know; when I was in London. I hadn't wanted him from the minute I knew about you. It was you I needed, really. Your belief in me, Jesus, Laura, you kept me *alive*.'

'I know,' I said. I felt the relief in her hand; the bones in it seemed to melt. It was time for straight talking, no more pussy-footing around. I refused to drag it out for her like he did for me.

'He set the fire and let me think it was you. I'm so sorry.'

There was a bewildered silence in which Beth's eyes darted over my face.

'Kit torched his own flat? With you in it? Kit?' Her expression made it clear: *I wouldn't have thought he'd have it in him*. I understood; I was still barely used to the idea myself. 'But *why?*'

'Because.' I could feel the tears coming but I pressed them back. Not because I needed to hide my grief in front of Beth but because I wanted to get as much of the story out while I was still coherent. I owed her that. 'He knew I'd never turn my back on you without a bloody good reason. So he gave me one.' I looked at my lap. 'He reckons you doing your friend's tyres and then that photo gave him the idea. You know, that you might be a bit . . .' I twirled my finger at my temple; for the first time it occurred to me that he might have somehow even orchestrated those things. Beth read my face and held up both hands.

'No, I'm afraid that was all me.'

I let that settle for a while. We all do stupid things when we're young.

We sat there in silence. In the corridor, a trolley clattered past, and from somewhere wafted a disgusting acrylamide whiff that could only mean hospital food. I picked a grape – not one of mine – and broke its skin between my teeth. Its juice was tart in my mouth.

'Why *didn't* you tell me about you and Kit?' It came out like an accusation. Maybe on some level that I couldn't police, that was how I really felt. My life sentence is to examine every prejudice in myself as well as others.

'I was going to,' she said, colouring a little. 'That's why I was there, in that field. During the eclipse. I'd gone after you – I'm not proud of it – I was going to tell you what a wanker you were going out with.'

'So what stopped you? Was it Jamie?'

'No, I was already leaving when he caught up with me. *You* stopped me.' My incomprehension must have shown; she tilted her head to the side, and her voice was almost a whisper. 'I watched you watching the eclipse. I saw you both together; it was magic.' Her words took me back to the top of the lorry, watching violet light stain the sky, so wrapped up in each other we might have been the only humans in the world. 'He'd told me the night before what eclipses meant to him and it was *still* an effort for him to take his eyes off you to look at the sky. I'd never had anyone look at me like that. Still haven't.' She twisted the corner of the blanket in her hands. 'I knew he'd been telling the truth, that what happened with me was a moment of madness, and everyone's got the right to fuck up, haven't they? You were *good* together, Laura, you were the real thing. It's something you recognise when you see. I couldn't touch it – oh, I'm sorry.'

My tears were hot and fast. Beth was saying the only thing I could no longer bear to hear. We had been good, me and Kit; we had been golden. For a blinding solar flash, I saw why he had gone to such vile lengths to preserve us. A yowling noise escaped my throat. Beth handed me a tissue from her bedside cabinet, and took one for herself. 'If we hadn't been thrown together by the court case, that would've been an end to it. I would've chalked the thing with Kit up to experience and left you both alone and all this . . . all this *shit* would never have happened.'

I escaped to this notional other world for a second. There, Kit would never have been tested, never have known his dark capabilities. The gold would have faded over time but never to this base metal. And my children would have grown up with their parents together.

Or perhaps, in this parallel life, getting away with fucking Beth would have given him a taste for it. Perhaps he would have spent the rest of his life making up for his perceived lost time. I'll never know. And in that other world . . .

'If we'd never gone to Cornwall, Jamie would still be alive,' I thought aloud.

'Well,' Beth looked straight at me. 'It wasn't a *completely* wasted journey, then.'

Our laughter was sick, uncontrollable and indivisible from our tears. There was a word for it: hysteria. The ancient Greeks believed that a woman's uterus roamed around her body making her mad. My uterus was anchored by babies to my pelvis, but as we sat in that hospital, rocking with wild laughter that had the nurses running in, I felt that some even more fundamental part of me had broken away and was running loose inside me.

Eventually the laughter died down and we sat in silence that was broken only by the occasional ragged sigh. I took Beth's hand again and we stayed like that, the curtains pulled tight around us. In the atmosphere of utter honesty, I came close to telling her that I had lied for her in the witness box; or rather, asking her if she *knew* that I had lied for her. Then an orderly wheeled in supper – mash, cabbage and something brown – and the spell was broken. So what if she knew I had lied? Beth and I had walked our separate hells and finally reached safety together.

I rattled the curtain rings on my way out of the cubicle. 'Will you come and see me again?' she asked. I turned around but didn't answer straight away. Kit's voice carried across the years, cutting me off. *How can you seriously expect to forge a genuine friendship with someone you met like that?* Beth's mouth quivered as if she was waiting for my permission to smile. I mentally silenced Kit. Let me try to get one good thing out of this mess of smoke and steel and lies.

'Of course,' I said.

I walked along the shiny corridor, aware of a loosening and release deep inside my chest; the hard guilty stone of my perjury crumbling into nothing. Unlike Kit, I had lied and got away with it. At this thought, something caught in my throat. No. It had not quite crumbled to nothing. I understood, as I left the hospital, that that rock of guilt would never dissolve entirely. I stood on the filthy street, cars rushing past me, as it dispersed into powder, quite invisible, and fine enough for me to carry in my veins for ever.

66

LAURA
30 September 2015

There are two photographers outside my house, shooting the cab even before they can know who's in it.

'Where's Christopher, Laura?' shouts one of them through the window.

'Want me to come in with you?' asks Ling.

'I'm good. You go back to your girls, they've barely seen you all week. Ring you later.' I keep my head down, and make it to the front door with my jacket pulled high over my head, my hair hidden. I gave them their kiss outside the court; clearly I was naive to think the story could end there. I force the smile it's so important to show my twins. They have already seen enough of my crying to scar them for ever. Those muslin cloths that you suddenly acquire by the dozen with newborn babies are perfect for mopping up tears. For a while I walked around with one on each shoulder; one for a baby, the other to cry on.

The tag on the doorbell beside me now simply reads *Langrishe*.

Beth opens the door to me, stepping backwards so that the photographers can't see her. She's wearing one of my old milk-stained maternity tops and a pair of shiny leggings. There is so much to say, but for now, 'Well done,' she says, drawing me into a deep hug. 'You got through it.'

'*We* got through it,' I correct her.

He Said/She Said

Beth's smile is watery; we'll do this later, over a drink. She came to visit the babies when they were two weeks old. This time, I begged her to stay. 'Why *her*?' asked Ling, when I hired her as my nanny. 'Doesn't she just remind you of all the shit you've been through?' To which I could only reply, 'Who else?' All I see now are the opportunities she *didn't* take to hurt me. Who could I trust more?

And anyway, how would I explain it all to a stranger?

'Fin's just woken up,' says Beth. 'I'm getting his bottle ready.'

'Perfect.' I kick off my shoes. In the sitting room, my oldest boy lies on his back, bashing the toys dangling from his activity bar like a manic little percussionist. I kneel to breathe in his almond smell and laugh as he grabs for my earrings. He's changed even since this morning; his saffron eyelashes are longer, or maybe he's just got more hair. He's got his dad's nose, his granddad's ears and an oval face that's all mine. He's a clown and a bruiser where Albie is soft and sensitive, watching intently for patterns and outcomes before committing to anything; Fin ploughs headlong into whatever's on offer. I try very hard not to compare them to Kit and Mac, and I may yet succeed.

'Have they been good?' I ask, disentangling Fin's fist from my hair.

'Albie's been an angel, Fin's been a little sod,' she says fondly.

'Did you talk to Antonia?'

'She's hiding in a neighbour's, watching the paps over the road,' she says.

'Oh, no,' I say, although I ought not to be surprised by this by now. Maybe I should text Kit and tell him to come in the back way, through Ronni's garden.

'He'll be here in about ninety minutes. I said he could help with bath time.'

Beth's face pinches in on itself. 'Want me to make myself scarce?'

Part of me wants to say no, stay, let's make him squirm his way through this. I'm sure it would be more convenient for Kit if Beth

401

suddenly disappeared for good now. That's not going to happen; but neither can she be here when he comes home. The way I manage his homecoming sets up the template for the rest of our lives. Kit has lost my love but he is still my children's father. 'Just this first time,' I say. 'Thank you.'

Once Kit is here, a new nightmare will slide into the place of the old. There will be more lawyers, I suppose; mediation to start with, issues of access, the house, custody – he'd be *insane* to contest me on this, but I can no more predict his actions now than I could those of a stranger in the street – and, when he eventually gets a job, possibly even child support. He might have something to say on the subject of his sons' surname. If he challenges me on my choice of childcare, he'll have a fight on his hands.

A soft grizzling noise comes down the stairs as Albie wakes up and realises his brother has gone.

'Want me to get him?' asks Beth, but it's a rare treat for me to wake either boy from his afternoon nap. I hand her Fin and climb to the nursery. The boys sleep in the attic room that used to be our study. Our desks and clutter have been replaced by two cots and a changing table, all from eBay. By the time they're old enough for real beds, I'll probably have had to sell up. It will do for now.

I tiptoe in even though Albie's awake, and pause for a moment on the threshold. I've had the huge plate window fitted with a blackout shutter and the only light comes from the illuminated globe that turns gently on its axis, painting the white walls with maps.

'Hello, sleepyhead,' I murmur into the satin of his neck. I throw back the shutters and the sky spills in, a wide expanse of blue above the rooftops. Together we look out over them. The sky over Alexandra Palace is bright blue, streaked with white, the baby-boy-nursery colours. Albie stares unfocused at scudding clouds until something catches his eyes and makes them widen in delight. A low-flying aeroplane booms its vapour trail slowly across the sky. Albie points, the first time he's made the gesture. I follow my son's gaze skywards, forwards. It's going to be so hard, from now on, but I won't look backwards any more, or over my shoulder. Albie knows this truth: we are meant to look up.

The new kitchen island is laid out with formula and sterilising equipment. Beth is preparing a bottle. Fin sits on the floor, more or less in the spot where she fell and bled out. It has never seemed to bother Beth, living and working in the place she nearly died. Perhaps because it no longer looks like the murder room. I had to change it. The house was a crime scene for three days. While the CSIs took photographs and dusted and measured, the blood was seeping into the porous tiles. Even the team of cleaners they sent could not bleach it out. I had the whole space refitted with end-of-range tiles, half-price Ikea units and catering equipment that Mac got on the cheap. It looks like an operating theatre now. Losing the kitchen's character is a small price to pay for losing the blood-stains too.

I can enter it now without flashbacks, but the dimensions are the same, and it's impossible for me not to superimpose the old layout on the new. There stood the old fridge where we kept all my fertility drugs. There remains the window where we stood in the sunlight, the sooner to read the positive pregnancy test. There was the worktop where he cooked me dinner every night. There in the window was the radio that soundtracked our kitchen discos.

There is a kitchen table and chairs in place of the old-fashioned banquette and counter where Jamie Balcombe held me and Beth hostage. There is the threshold where Kit stood and watched, paralysed by what he saw and what I knew. And here, just to the right of the dishwasher, is where I turned my husband into a killer.

The 'mad scramble' for the knife we both described in court was nothing of the sort, although only I knew this.

I knew that Jamie had to die if we were all to live.

The sketch I had of what Kit had done sent me wild with fury. Let the blood be on your hands, I thought. I will not give birth in prison. I will not give up a second's liberty for you.

All those years playing pool taught me about trajectory and precision, and I also learned how to bluff. I knew exactly what I was doing as my fingertips bounced the knife Kit's way, and I watched with satisfaction as his faithless hand closed around the hilt.

Acknowledgements

Thanks to my fabulous new editor Ruth (I have thoroughly enjoyed my first Trossing), Louise Swannell, Leni Lawrence, Cicely Aspinall, Naomi Berwin, Penny Isaac, and all at Hodder and Stoughton.

To the wonderful team at United Agents: Sarah Ballard and Zoe Ross, Margaret Halton, Amy Mitchell, Joey Hornsby, Eli Keren and Georgina Gordon-Smith.

My learned friends: Daniel Murray, Bathsheba Cassell, Harriet Tyce, Gemma Cole and Chris Law. Special thanks are due to the author and lawyer Neil White, who answered a tweet asking for 'five minutes' of his time and was still fielding daily emails six months later. Any legal or procedural mistakes are mine, and there will be no further questions.

To the coven: Mel McGrath, Louise Millar, Jane Casey, Laura Wilson, Kate Rhodes, Sarah Hilary, Serena Mackesy, Helen Smith, Denise Meredith, Ali Turner, Alison Joseph, Katie Medina, Helen Giltrow, Louise Voss, Colette McBeth, Paula Hawkins, Tammy Cohen and Nikoline Nordfred Eriksen.

Thank you Helen Treacy and her indefatigable red pen, Sali Hughes for calling BS on hippies, and Julia Crouch and Claire McGowan for always knowing whether I need wine or tea.

Finally, and above all, thank you to my family: Mum and Jude, Dad and Susan, Owen and Shona. And to Michael, Marnie and Sadie. I love you.

I'd like to acknowledge the following resources:

Carnal Knowledge: Rape On Trial by Sue Lees

Eve Was Framed: Women and British Justice by Helena Kennedy, QC

Total Addiction: The Life of An Eclipse Chaser by Dr Kate Russo

Totality: Eclipses of the Sun by Fred Espenak, Mark Littman and Ken Wilcox

'*Dancing in the Cosmic Sweet Spot*' blog: Graham St John